T0292358

# The Care of
# Natural Monuments

# The Care of
# Natural Monuments

## with special reference to
# Great Britain and Germany

by

## H. CONWENTZ
Prussian State Commissioner for the Care of Natural Monuments

With ten illustrations

CAMBRIDGE:
at the University Press
1909

# CAMBRIDGE
## UNIVERSITY PRESS

University Printing House, Cambridge CB2 8BS, United Kingdom

Cambridge University Press is part of the University of Cambridge.

It furthers the University's mission by disseminating knowledge in the pursuit of education, learning and research at the highest international levels of excellence.

www.cambridge.org
Information on this title: www.cambridge.org/9781107433274

© Cambridge University Press 1909

This publication is in copyright. Subject to statutory exception and to the provisions of relevant collective licensing agreements, no reproduction of any part may take place without the written permission of Cambridge University Press.

First published 1909
First paperback edition 2014

A catalogue record for this publication is available from the British Library

ISBN 978-1-107-43327-4 Paperback

Cambridge University Press has no responsibility for the persistence or accuracy of URLs for external or third-party internet websites referred to in this publication, and does not guarantee that any content on such websites is, or will remain, accurate or appropriate.

TO

THE RIGHT HONOURABLE

RICHARD BURDON HALDANE

K.C., LL.D., M.P.

SECRETARY OF STATE FOR WAR

IN REMEMBRANCE OF OLD TIMES
SPENT TOGETHER AT THE UNIVERSITY

TO

THE RIGHT HONOURABLE

RICHARD BURDON HALDANE

K.C., LL.D., M.P.

SECRETARY OF STATE FOR WAR

IN REMEMBRANCE OF OLD TIMES

SPENT TOGETHER AT THE UNIVERSITY

# PREFACE

ON August 1st, 1907, I delivered a lecture, on the Care of Natural Monuments, at the Leicester meeting of the British Association for the Advancement of Science. The lecture was given at a joint meeting of the Geographical, Geological, Botanical, and Zoological sections.

The present book may be regarded as the above lecture in an extended form ; and, in publishing it, I am acceding to the expressed wishes of my English friends.

Whilst the book makes no pretence to deal exhaustively with the subject, I hope that it is sufficiently comprehensive to be acceptable and serviceable to naturalists, and, indeed, to lovers of nature in general.

There remains the pleasant duty of thanking the various British authorities, societies, and naturalists who have generously supplied me with information. In particular, I beg to tender my heartfelt thanks to Professor F. W. Oliver, Mr Clement Reid, Professor A. C. Seward, Dr W. G. Smith, Mr A. G. Tansley, and, last but not least, Dr C. E. Moss, who has kindly revised my English manuscript, and seen my book through the press.

H. C.

30 *November*, 1908.
DANZIG.

# CONTENTS

x    *Contents*

# ILLUSTRATIONS

# INTRODUCTION

The expression "natural monument" or "monument of nature," from the German "Naturdenkmal," is new to the English language, and it is necessary, therefore, to explain its meaning. First, what is meant by the single word "monument" or "Denkmal" must be defined. The term is usually applied to anything established in commemoration: there are monuments erected in remembrance of eminent persons, such as Darwin, Nelson, Scott, and Shakespeare, of brave deeds, and of famous incidents. Further, standard works of literature and music may be called monuments of literature. Ancient buildings, such as cathedrals, castles, and monasteries which have a historic or an artistic value, are spoken of as monuments of architecture and of art,—in German "Bau- und Kunst-denkmäler." The expression is also applied to prehistoric remains, such as lake dwellings, stone circles, dolmens, and burying mounds which are spoken of as prehistoric monuments.

All the above-mentioned objects are the result of man's activity ; but nature too has her monuments. Just as the ornamented stone obelisk is a monument of art (Kunstdenkmal), and as the rude stone, erected by man in former ages to the memory of the dead, is a prehistoric monument, so too the erratic block, transported in past time by natural forces, constitutes a natural monument. Again, just as an artificially erected wall or mound of an earlier period may be a prehistoric monument, so the natural moraine or prominent hill produced without the help of man is a monument of nature Moreover, examples of beautiful scenery, characteristic soil-formations, interesting associations of plants and animals, and rare species of the indigenous flora and fauna may all be classed as monuments of nature. A virgin soil-formation, however, is not often to be found in densely populated countries, most portions of which have been under cultivation for many centuries. For that reason the term "natural monument" is occasionally used in a broader sense.

It may also be mentioned that old big trees had been spoken of as "monuments de la nature" by Humboldt[1] ninety years ago, but latterly the term has fallen into disuse.

[1] Humboldt, A. de. Voyage aux régions équinoxiales. Tome II. Paris, 1819, p. 59.

# NATURE THREATENED

Anyone travelling in England and Germany, or, in fact, in most parts of the world, may see for himself that the constant cultivation of the land and the growth of industrial undertakings have threatened, and in many places considerably damaged, interesting tracts of country as well as natural monuments. From the economic aspect, this is immaterial; and it would even be justifiable, from such a point of view, if man were to bring under his control almost the whole realm of nature. But, on the other hand, from the scientific and aesthetic standpoints, it is much to be regretted that so many types of scenery and of the vegetable and animal worlds should pass away irrevocably. In his book " Unto this

Last" John Ruskin says: "As the art of life is learned, it will be found at last that all lovely things are also necessary; the wild flower by the wayside, as well as the tended corn, and the wild birds and creatures of the forest, as well as the tended cattle; because man doth not live by bread only..."

## Views

There are many beautiful hills and mountains on which buildings of various kinds have been erected to the disfigurement of the neighbouring scenery. Surrounded as we are in large towns by man's works, in the mountains, at least, we might be allowed the enjoyment of nature in its entirety.

Again, certain famous views in Germany, Switzerland, and other countries are crossed by railways, which disfigure the scenery, and are, in many cases, quite unnecessary. It was proposed to construct funicular railways and lifts to the Hexentanz-platz and the Rosstrappe in the Hartz mountains, as well as to the Bastei in Saxony; but the Prussian and Saxon government refused to entertain any such proposals, in order to preserve those famous view points in their original state. Again, a beautiful valley in Thuringia, the Schwarzatal, was threatened with a railway which was to have been constructed along the valley. The government of Schwarzburg-Rudolstadt, however, refused the necessary permission, on the ground that this magnificent natural scenery ought to be preserved invio-late. Hence the railway now follows a more circuitous route.

## Water

The utilisation of water power generally yields a fair profit. Falls and rapids in particular are utilised by industrial undertakings, as on the Tivoli Falls in Italy, on the Rhine Fall in Germany, and on the Niagara Falls in America. Perhaps no other place of natural beauty in the whole world is so utterly disfigured as the Trollhätta Falls in Sweden. These Falls are now surrounded by numerous manufactories such as electrical works, engine works, an iron foundry, an oil factory, cellulose works, and carbide works. Moreover the opposing rock is disfigured by glaring advertisements. Hence this beautiful monument of nature has become her caricature. The Swedish government endeavoured to buy out these establishments, but to no purpose. In the year 1899, a

Waterfall Committee for the registration and charting of all the public falls and rapids in Sweden was instituted by Parliament. In the report of the Committee, it was proposed not to utilise every one of the falls for industrial purposes; but to reserve, from the beginning, a few falls or rapids as natural monuments[1].

## ROCKS

Then, again, remarkable rocky districts are frequently disfigured by stone-quarries. On the western and eastern coast of southern Sweden, granite is quarried for paving-stones, which are exported chiefly to the continent. On some parts of the coast, the quarries are rather extensive,

[1] Betänkande afgifvet den 17. mars 1903 af den för utredning beträffande vissa staten tillhöriga vattenfall af kungl. Majestät den 9. juni 1899 tillsatta kommittén. Stockholm, 1903, p. 117.

as for instance between Karlshamn and Karlskrona, a distance of about thirty miles. It is, no doubt, owing to this fact that the whole stretch of coast has been changed, and even scientific rarities have been destroyed.

In England, the once beautiful Miller's Dale, in Derbyshire, and the famous gorge of Cheddar, in Somerset, are similarly disfigured (C. E. Moss).

In Saxon Switzerland, on both sides of the Elbe, there are nearly 300 quarries, owing to which some examples of the most beautiful scenery in Europe have been destroyed. Between the Bastei and Pirna, the quarrying of sandstone takes up more than half the bank of the river.

Of course, the stone industry yields a very good return. From it, Sweden obtained in 1901 an income of more than £500,000, and Saxony more than

£200,000. Nobody expects that such a profitable industry should be stopped solely for the preservation of natural monuments; but it is highly desirable that here and there its sphere of action should be restricted. The Saxon government has recently determined not to lease quarries on the bank of the Elbe, and also not to establish new ones there again.

Not only the rocks of the mountains, but also the boulders of the plains are threatened by industry. The larger ones are worked into pillars and other objects, and the smaller ones into paving stones. Thus, many low lying districts are already deprived of their erratic rocks. The Sarsen Stones on the Marlborough Downs, well known as the Grey Wethers, were thus endangered, in consequence of a recent change of ownership. Steps were therefore taken and an appeal was formulated to raise the money for preserving some characteristic specimens of the stones in their natural condition (see page 84).

## FENS

Fen districts are often characterised by deposits of peat, the remains of which possess a resemblance to existing plants and animals, and also contain subfossil remains, which in some cases reach back to the glacial period. In former times, even so late as the seventeenth century, peat was considered of no value. The Dutchman Pickardt once said that peat was, by the Lord's punishing hand, created for the vexation of mankind. Since then, public opinion with regard to peat has undergone an entire change. These fenny districts have been drained more and more, and have been utilised for agriculture and industry, so that in many civilised States, there is now no primitive fenland at all. From the standpoints both of industry and health, this state of things is admirable; but from the scientific one at least, it is deplorable that the vegetation linking the present to the past should be destroyed[1] before scientists have become fully acquainted with it. It would be absurd to expect that no portion of the fens should be used; but no country is so poor that it cannot preserve one or two tracts of fen for the purpose of study and education. In Great Britain, strips of Wicken Fen and Burwell Fen in Cambridgeshire are safeguarded by the National Trust (see pages 75 and 84). In Germany also

[1] Conwentz, H. Die Gefährdung der Flora der Moore. *Prometheus*, XIII. Jahrgang. Berlin, 1902, S. 161.

there are several fens protected in a like manner.

## WOODS

By the term wood is here meant a natural or semi-natural association of woody and herbaceous plants. These are to a great extent threatened by cultivation. Particularly, clear felling, now resorted to in many localities, is destroying not only the dominant forest trees, but also the associated trees and shrubs, such as (to quote but a few instances) *Taxus baccata, Juniperus communis, Salix caprea, S. aurita, Populus tremula, Corylus Avellana, Ulmus montana, Prunus Padus, P. spinosa, Pyrus communis, P. Malus, P. torminalis, P. Aucuparia, Euonymus europaeus, Ilex Aquifolium, Rhamnus cathartica, Rh.*

*Frangula*, *Tilia parvifolia*, *Fraxinus
excelsior*, *Cornus sanguinea*, *Sambucus
nigra*, *Viburnum Opulus*, and *Lonicera
Periclymenum*. The herbaceous plants,
too, and the ferns and mosses are
more or less destroyed; and at the
same time the birds, which live and
breed in the woods, are disappearing.
In place of the old natural woodland, a mere plantation arises which
consists only of cultivated and perhaps
foreign species of forest trees. In this
way, the elements of the forest, the
vegetable and animal world, and indeed the whole aspect of the country-
side are greatly altered in character.

In many countries, for example, in
Holland, Denmark, and Saxony, the
primitive woods have largely disappeared. Whilst travelling in Great
Britain, I have occasionally seen places
where trees are to be found in their
natural growth, but I am not sure that

there is any large wood which can be said to be wholly indigenous.

With regard to France, G. Andersson reports on the devastation of forest in the Cevennes, where (he states) only box scrub remains[1].

Engler also gives an account of the "falsification" of the original character of the vegetation of Table Mountain above Cape Town, through the introduction of Mediterranean conifers[2]. Moreover it is well known to what an extent the woods have been cut down in the United States of America. It is worthy of remark that a woodman felling a Big Tree stands for the coat of arms of one of the States.

---

[1] Andersson, Gunnar. Skogsköfling och skogs-odling i Cevennerna. *Skogsvördsföreningens Tid-skrift.* Stockholm, 1903, p. 213 sq.

[2] Engler, A. Über die Frühlingsflora des Tafelberges bei Kapstadt. *Notizblatt des König-lichen Botanischen Gartens.* Appendix XI. Berlin, 1903, S. 23.

## PLANTS

Many plants are threatened with extinction, some by thoughtlessness and others by trade. Every Sunday, crowds return from excursions bearing great bundles of flowers destined to wither in a few hours, or even to be thrown carelessly by the roadside. People have not yet learned that nature can be best appreciated in her secluded haunts, and that her treasures are too sacred to be carried into the withering atmosphere of the city. They should always remember that, whilst thousands may look at a beautiful plant, only one may pluck it. Some species which are especially subject to extirpation may be mentioned here.

The parsley fern (*Cryptogramme crispa*) is frequently removed in great quantities from its native haunts in the Lake District by some of the florists of the towns of Lancashire and Yorkshire (C. E. Moss).

The yew (*Taxus baccata*) formerly occurred all over Europe; but its excellent timber was in much request for bows and other objects, particularly in Great Britain. It is well known that in former times enormous quantities of its timber were exported to England and Scotland from Germany, Austria, Russia, and Switzerland. Even at the present time, a Royal Bowmaker is living in Edinburgh; and, some years ago, I spoke to him about the yew wood which he handled. He had then only one large piece from the Caucasus, and did not know where he could obtain more. This demand for the timber of the yew has contributed largely to its disappearance. Of course, in many parts of Great Britain and Ireland, the species still occurs. Still, the submarine woods on the English coast, and other subfossil records, and the numerous prehistoric and historic specimens of yew, which I have seen in the Science and Arts Museum at Dublin and in other museums in Britain, show that formerly the yew was more common. In Sweden, the tree is, generally speaking, not rare; but when I was living at Stockholm, in the autumn of 1897, I saw every day, for two months, a great many branches brought over from the islands for sale at the market. In Denmark, the yew still occurs in one place, near Vejle, where it was discovered by English engineers during the construction of a railway. But the National Museum of Copenhagen and other Danish museums contain many prehistoric objects made of yew which for the most part at least were worked there. In Holland, and in several districts of Germany, for example, in the provinces of Schleswig Holstein, Brandenburg, and Posen, the tree is quite extinct.

The lily of the valley (*Convallaria majalis*) is also endangered by dealers. In a village near the Vistula, a man applied for leave of absence from school for his eleven children in order that they might gather the lily of the valley in a neighbòuring wood. In the isle of Rügen, during the spring of 1898, 3,400 kg of this plant were gathered, and exported chiefly to Berlin. In this way the plant is becoming rare here and there, and thus it is that a charming memento of the woods is dying out. In certain of the Derbyshire dales, in England, the lily of the valley has been almost exterminated.

The Lady's slipper orchid (*Cypripedium Calceolus*), one of the most beautiful members of the European

flora, is everywhere threatened. It
is dug up by gardeners and florists.
In England, the plant (always local)
is now very rare, and exists in only a
few localities. Forty years ago, it was
said of it: "Once plentiful in Castle
Eden Dene, but now, we fear, nearly
extirpated[1]." The species, however,
is not quite extinct in England, as
Mr J. G. Baker wrote to me in August,
1906:—"It still lingers at the old
locality in Durham (Castle Eden Dene)
and was gathered last year at two
places in Yorkshire..." In all Denmark, it has only one habitat (see
page 104). On the chalk rocks of
Thuringia, the plant is still fairly common, although even there it is threatened by trade. Near Jena, several

---

[1] Baker, J. G., and Tate, G. R. A new Flora
of Northumberland and Durham. *Natural History
Transactions of Northumberland and Durham*,
Vol. II. London, 1868, p. 256.

years ago, a man was arrested who had collected about 700 specimens. Though he had not previously been convicted, he received a sentence of a fortnight's imprisonment. The judge remarked : " The Civil Court has been obliged to have recourse to this somewhat drastic punishment, because, in the neighbourhood of Jena, where it is well known that the most rare and beautiful orchids occur, the unscrupulous uprooting of these rare plants has become such a nuisance that unless severe measures are taken to prevent this practice, the time is not far distant when the most beautiful ornaments of our Flora will have been exterminated."

At Munich, I saw one day in 1902 on the open market, seventeen various species of wild mountain plants, and bunches of six to ten plants of the Lady's slipper were to be had for a penny ! In the kingdom of Saxony, the species has already been entirely destroyed.

The dwarf-palm (*Chamaerops humilis*) used to grow near Nice, which was the northern limit of this, our only European palm. It is constantly being uprooted there, however, by gardeners and builders. Thus, the limit has considerably receded towards the south ; and the species has almost vanished as a member of the Central European Flora (Ascherson and Graebner).

The existence of even a common tree may be threatened by a great demand for its timber. In Sweden, the aspen (*Populus tremula*) is disappearing, as its timber has been largely used up in the production of matches. At present, aspen timber is imported for that purpose from Russia.

The English custom of decorating their homes with mistletoe (*Viscum album*) at Christmas demands large quantities of this hemi-parasite. For

instance, during the Christmas season
of 1907, 380,000 kg were sent from
Brittany to England.   But this custom
is very popular not only in Great
Britain, but also in German towns,
wherever English colonies exist, as,
for example, in Berlin, Dresden, and
Munich.   Moreover, in Germany this
custom is gaining ground more and
more.   In Berlin, at Christmas time,
there are enormous quantities of mistletoe for sale, both of the proper species
(*Viscum album*) and also of the variety
(*Viscum album* var. *laxum*) which grows
upon the Scots pine.   As a natural
consequence of this trade, both plants
are becoming scarce in many of their
native haunts.

In England, the Cheddar pink
(*Dianthus caesius*) is frequently exposed for sale by the cottagers at the foot
of the gorge which furnishes its only
British station (C. E. Moss).

Generally speaking, in the neighbourhood of towns, plants with showy and fragrant flowers are sacrificed to money-making. Thus, the flora of Vienna was formerly distinguished by many remarkable species, such as *Lilium Martagon*, *Muscari racemosum*, *Scilla bifolia*, *Iris germanica*, *Cypripedium Calceolus*, *Ophrys muscifera*, *Orchis* spp., *Helleborus viridis*, *Dictamnus alba*, *Daphne cneorum*, *Primula Auricula*, and *Gentiana verna*. At present, these interesting plants are being greatly reduced in numbers, whilst some have been totally extirpated (v. Wettstein).

Other examples of rare plants endangered by cultivation, especially in England, are published in some papers of G. S. Boulger, one of which is mentioned below[1]. At the Leicester

[1] Boulger, G. S. The preservation of our wild plants. *Journal of the Royal Horticultural Society*, Vol. XXIX. Part 4. London, 1905.

meeting of the British Association, in Section K, Mr A. R. Horwood, of Leicester, read a paper on the disappearance of cryptogamic plants[1].

## ANIMALS

The animal world is ultimately dependent on vegetation, so that if a bog is drained or a wood cut down, a great number of animals, as well as plants, are destroyed. It must not be supposed that the animals return after a new plantation has been formed on the old woodland site, for many species live only where there is a natural plant growth and not on cultivated land. In Germany, for instance, about 80 species of spiders live on cultivated ground, in fields and forests, and by

[1] *British Association Report*, 1907. See also *Irish Naturalist*, Aug. 1908.

road sides; but more than 500 species live only on peat, on heaths, amongst woods, and in other natural habitats (F. Dahl).

Birds afford a great many articles of trade, such as plumes, skins, and eggs; and this circumstance contributes largely to the fact that some species are disappearing, whilst others are already extinct. At Nice, in the market-hall, 1,318,356 singing birds and other small species were exposed for public sale during the three months Nov. 1st 1881 to February 1st 1882. Humming birds and birds of paradise are also diminishing. The South African ostrich (*Struthio australis*) has been exterminated in Cape Colony, in Caffraria, and in Natal. A kind of parrot (*Conurus carolinensis*), formerly extensively known in the United States of America, is at present only to be found in the most southern districts.

The magnificent Marno of Hawaii (*Drepanis pacifica*) was destroyed about fifty years ago.

Again, the existence of certain mammals living in the water or on land is increasingly threatened by trade. Ferdinand v. Richthofen says[1]: Though the cruel lust for slaughter has long ago succeeded in reducing the numbers of marine mammalia to an alarming extent, and has even utterly exterminated some of the noble species, yet the task of fishing near the surface of the ocean has remained a laborious and dangerous one. The trawl—originally an instrument of capture used by zoologists alone—has grown to its present gigantic proportions, and is even used for deep-sea fishing. This systematically sweeps

[1] Richthofen, Ferd. Frhr. v. Das Meer und die Kunde vom Meer. *Akademische Festrede*. Berlin, 1904, S. 31.

the bottom of the sea like an enormous steam plough, and thus completes the destruction begun by the large sweep nets used nearer the surface. Thus there has arisen an imperative need for a rational regulation of the exploitation of marine life. Indeed, this is felt to be quite as urgent a necessity as in the case of the magnificent mammals of Africa, the fur-bearing animals of Siberia and Canada, and the gaily coloured birds of the tropical forests. For, is it not true that an exterminated species of animal or plant is gone for ever?

The beaver (*Castor fiber*) formerly occurred over the whole of Europe except the extreme north. This is clear from fossil finds, and from names of rivers, lakes, and bogs. In Germany, for instance, I know more than 300 localities whose names are derived from that of the beaver. In Sweden, the town of Hörnesand, half-way up the Baltic coast, shows a beaver in its coat of arms. Recently the animal has become very rare, generally. For instance, in the Vistula, the last specimen disappeared in the middle of the nineteenth century, and the last beaver from the Danube was to be seen alive in the Universal Exhibition at Vienna, 1873. At present, in the whole of Europe, the animal is only known in the Elbe district in Germany, in the southern Rhone district in France, in a small district in the south of Norway, and in some places in Russia. In Germany and Norway, the beaver has come under the protection of law.

The American beaver (*Castor canadensis*) too, formerly the symbol of Canada, is quickly disappearing. The following figures give some idea of the number of beavers exported by a trading company :—

In the year 1887...102,715
,,      ,,   1895... 51,028
,,      ,,   1896... 49,613
,,      ,,   1897... 42,427
,,      ,,   1898... 36,402

Thus in twelve years the number was reduced to about one-third.

The fate of the musk-ox (*Ovibos moschatus*) is of even greater interest. In the glacial period, this animal was to be found in England as well as in Germany; and it still exists in Northern Canada and in Greenland. It will be seen from the following figures that the existence of the musk-ox is greatly endangered by the same company. The following figures indicate the number exported by that company :—

In the year 1891...1,358
,,        ,,   1894...  748
,,        ,,   1895...  473

In the year 1898... 449

  ,,         ,, 1899... 511

  ,,         ,, 1901... 271

Thus, during eleven years, the number of animals exported has been reduced to about one-fifth; and, at this rate, it is obvious that this rare and ancient species will in a short time become quite extinct.

Not only the Canadian beaver and the musk-ox, but also a great number of other animals are threatened by trade in those parts[1], and will become extinct if the methods of the traders are not, in the immediate future, seriously restricted.

The sportsman also is responsible for the extermination of many kinds of animals. For instance, the cormorant (*Graculus carbo*) was formerly common both in Germany and Great

[1] Nathorst, A. G. Hafva djuren rättighet att lefva? Stockholm, 1907.

Britain; but a veritable war of extermination has been waged upon this interesting swimming bird, which nests at the top of high trees. At Stegen, near Danzig, in the year 1862, in one day, 61 birds were killed by sportsmen. In another seaport town, an ornithological society, which should aim at the protection of birds, demolished the last nest of the cormorant near the town, by the aid of a fireman! At present, in all Germany, only one breeding-place of the bird is known. Further, the reindeer (*Cervus tarandus*), now confined chiefly to Spitzbergen, Nova Zembla, Greenland, and Siberia, is wantonly destroyed in the name of sport. In 1891–92, a well-known personage of high position, accompanied by ladies and gentlemen, travelled in Spitzbergen. In the Kolbay and Adventbay, more than 60 head were killed in three days: a few were utilised as food; but most of them were killed merely for the sake of the trophies which their antlers afforded. This regrettable case, I am sorry to say, is not the only one of its kind. Hence it is that the Spitzbergen reindeer is gradually disappearing; and this is the more to be deplored as it is a peculiar variety of the animal (*Cervus tarandus* var. *spitzbergensis*).

The African fauna, too, is rapidly diminishing before the unrestrained attacks of man. Sir H. H. Johnston and Professor C. G. Schillings are quite right in raising a protest against the idea that the destruction of the animal world is part of a fashionable man's education, and against the enormous damage done by obscure sportsmen[1].

[1] Schillings, C. G. With Flashlight and Rifle. Translated by F. Whyte. With an introduction

Without doubt, these kinds of hunting are not at all sportsmanlike, for an old German saying reads thus :

> "Das ist des Jägers Ehrenschild,
> Dass er *beschützt* und *hegt* sein Wild,
> Waidmännisch jagt wie sich's gehört,
> Den Schöpfer im Geschöpfe ehrt."
>
>                 (RIESENTHAL.)

by Sir H. H. Johnston. With 302 illustrations. London, 1906.

# NATURE PROTECTED

The above mentioned instances, which could easily be added to, serve to show that many monuments of nature are far advanced towards ruin, whilst others have been already destroyed. Everywhere people are becoming loud in their protestations; and a widespread feeling has arisen that as much as possible should be done to prevent further destruction. The opinion is growing that, in general, monuments of nature as well as of art should be guarded against irretrievable ruin wrought by man. In England, Germany, and other States various measures have already been taken for this purpose. Before discussing these measures, the aims of the movement for the care of natural monuments may be stated.

## Aims

In general, there are three aims to which these efforts should be directed: first, to establish a register or inventory and to map the natural monuments; secondly, to preserve them *in loco*; and thirdly, to publish accounts of them.

## Registration

For a long time past, an inventory of monuments of architecture and art has been taken in nearly all civilised States[1]. Monuments of nature, so far as they are known, should now

[1] Wussow, A. v. Die Erhaltung der Denkmäler in den Kulturstaaten der Gegenwart. Zwei Bände. Berlin, 1885.

Brown, G. Baldwin. The Care of Ancient Monuments. An account of the legislative and other measures adopted in European countries for protecting ancient monuments. Cambridge, 1905.

be dealt with in a similar way. In
Germany, the soil, and the fossil and
recent vegetable and animal worlds
have not everywhere been thoroughly
investigated ; therefore these investi-
gations must be made anew here and
there. In these inventories, the geographical, geological, botanical, and
zoological rarities ought to be registered, not in systematic order, but
according to the administrative districts
and the ownership of the different
localities. Thus, from the inventory,
it would be easy to determine what
monuments of nature exist and ought
to be preserved in a given district.
Moreover, it is highly desirable that
reproductions of some kind should be
made, especially of those monuments
of nature which it is not possible to
preserve intact.

Further, the natural monuments
should be mapped out. Many natural

objects of great value and interest are
already localised in topographical and
geological maps issued by all civilised
States. As regards the botanical world,
the maps may be constructed in two
ways—on the one hand as vegetation
maps, and on the other as species
charts. The leading idea of the first
kind of map is that plant associations
or vegetation units should be studied
and charted. Such maps lead to a
complete knowledge of plant life in its
natural environment. They are not
primarily concerned with the precise
limits of individual species, but show
the distribution of selected plants associated in nature. One of the first and
most complete vegetation maps is that
of a portion of France, by Professor
Flahault[1], and it is based essentially

[1] Flahault, Charles.   Essai d'une Carte Botani-
que et Forestière de la France. *Ann. de Geog.*,
No. 28, 1897, pp. 289–312 (with map).

on the distribution of certain trees.
Vegetation maps are also being published in Great Britain (see pages 55
*et seq.*), Germany, the United States,
and Switzerland.

The second kind, species charts,
shows the actual distribution of a single
species. On these charts, either the
different habitats of a rare species are
marked, or the actual geographical
limits of its range of growth. For
example, figure 1 shows the distribution of *Gentiana lutea* in Württemberg
and the neighbourhood[1]. In figure 2,
the distribution of the spruce fir
(*Picea excelsa*), on the limit of the
geographical range in West Prussia,
is charted[2].

[1] Eichler, J., Gradmann, R., und Meigen, W.
Ergebnisse der pflanzengeographischen Durch-
forschung von Württemberg, Baden und Hohen-
zollern. *Beilage zu Jahresheften des Vereins für
Vaterländische Naturkunde.* 63. Jahrg. Stuttgart,
1907, S. 135 ff., Karte 6.
[2] Conwentz, H. Forstbotanisches Merkbuch,

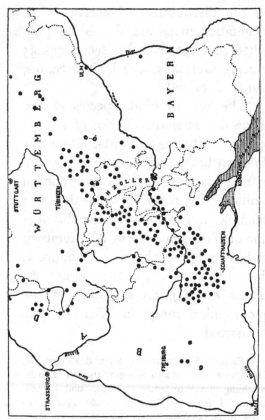

Figure 1. Chart of the stations of *Gentiana lutea* in Württemberg.

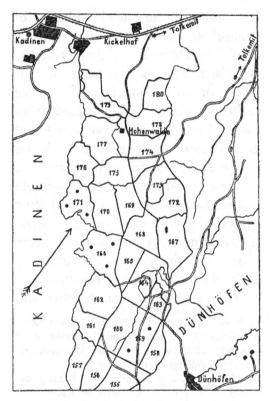

Figure 2. Portion of a Prussian forest-map, showing the [•] geographical limit of the Spruce Fir (*Picea excelsa*).

Similarly the nesting-places of remarkable birds and other animals should be charted.   Figure 3 shows the distribution of the nests of four species of birds in Switzerland[1].  These various maps may be revised from time to time, if such a course is found to be desirable.   The maps would be of great value for comparison, in showing to what extent the vegetable and animal worlds have changed after a lapse of twenty, fifty, or more years, in indicating what circumstances have caused the changes, and in assisting us in the care of what remains.

All these inventories, statistics, and

Nachweis der beachtenswerten und zu schützenden urwüchsigen Sträucher, Bäume und Bestände im Königreich Preussen. I. Provinz Westpreussen. Mit 22 Abbildungen.  Berlin, 1900, S. 5.

[1] Studer, Th., und Fatio, V.   Katalog der Schweizerischen Vögel bearbeitet im Auftrag des eidgen. Departements für Industrie und Land-wirthschaft.  I. Lieferung : Tagraubvögel.  Bern, 1889, Karte VII.

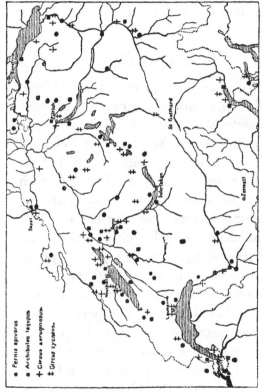

- Pernis apivorus
- Archibuteo lagopus
+ Circus aeruginosus
‡ Circus cyaneus

Basel · Zürich · St Gotthard · Interlaken · Zermatt · Neuchâtel · Lausanne

Figure 3. Chart showing the nesting-places of four birds of Switzerland.

maps ought to be collected in some central or local office.


## Protection

It has been proposed to institute National Parks in Germany and other countries on the lines of those in the United States (see pages 165 *et seq.*). It is, however, quite impossible to make reservations of land of such considerable area in any part of central Europe ; and further it is not sufficient to make only a few such reservations, even if they be large, in a whole country. Rather should numerous small reservations of land be effected in different parts of the country as more suited to the purpose in view. Here, for instance, the face of a cliff might be protected, and there a moraine or a boulder ; here a fall or rapid, and there an old river-bed with rare and interesting species, which perhaps are becoming extinct; in one district a stretch of meadow, and in another a tract of fenland; here a heath, and there a sand-hill with a pontic plant association; here a pine-wood, and there a deciduous forest.

Moreover, it has been proposed that governments should purchase all lands bearing, or constituting in themselves, monuments of nature. This proposal, however, cannot be carried out, as many places of that kind are not for sale. Besides, it is not always to the purpose that all natural monuments indiscriminately should come into one hand. In fact, it is only in those cases where the owner cannot be interested in the natural monument, and persuaded to protect it, that it is desirable for the State to purchase it. The required sums should be sub-scribed by public bodies, societies, wealthy citizens, and the purchased land should be given to some public body. No doubt, rich men who subscribe towards the acquisition of works of art would also be willing to contribute towards the preservation of the works of nature. Memorials should sometimes take the form of beautiful scenery, as has indeed already happened in the U.S.A. and to some extent in Great Britain. Such a memorial would be more lasting and more beautiful than the most perfectly executed monument in stone.

If possible, natural monuments ought to be preserved without changing surrounding nature. To this end, for instance, a natural history society in Germany made a grant for purchasing a large boulder, only on the understanding that it should not be enclosed by a fence. However, other cases may occur where it is advis-

Figure 4.   Marking of a rare tree (*Pyrus torminalis*)
in West Prussia.

able to hedge round the natural monument. Thus, if a rare tree in the forest is to be protected, it may be marked by little stones put round it. Figure 4 shows a Service Tree (*Pyrus torminalis*) marked in such a manner by the proprietress of an estate in West Prussia. The beauty of the forest is not impaired by these white stones : the woodcutter's attention is drawn to the tree ; and consequently he does not cut it down.

## Publication

It is true that in several cases a natural monument, the habitat of a rare plant or the dwelling-place of a notable bird, for instance, is best preserved by its not being published. It happened not long ago that after the publication of a map showing the whereabouts of the nests of certain birds, that natural history dealers went there for collecting eggs. From this it will be seen that it is inadvisable, in certain cases, that such maps should be published. As a rule, however, if people are to enjoy natural monuments, it is necessary to teach them how to read nature. For that reason, in schools, associations, and clubs the care of natural monuments should be specially taught and discussed. Again, the inventories of natural monuments of distinct districts should be published. The maps, for instance, the forest-maps with natural monuments marked upon them, ought to be printed and distributed to the forest officials, in order that they may always keep the rarities in mind. In a similar way, geological, zoological, and other inventories should be published.

### Realisation

The main point is, how are these ideas to be carried out ? I am happy to say that in several countries a beginning has been made in the registering and mapping out, as well as the preserving of natural monuments. With regard to protection, there are three general ways : by voluntary, by administrative, and by legislative help. In the remaining pages, a number of precedents of each kind are mentioned for various countries which are alphabetically arranged.

### Austria

In several provinces of the Empire, laws exist for the protection of remarkable plants, such as *Pinus pumilio, Cypripedium Calceolus, Ophrys muscifera, Daphne Blagayana, Primula*

*auricula, Gnaphalium Leontopodium*
(Edelweiss). The last mentioned plant
is protected in all provinces where it
occurs. In 1902, the Minister of Education was enabled by Parliament to
promote the care of natural monuments. He requested from the philo-
sophical faculty of the university of
Vienna a report on the value of these
endeavours. The reports treat of the
protection of natural objects of geographical, geological, botanical, zoological, and aesthetic interest. For the
future, the state offices were to be
charged with the registering of the
natural monuments in their districts.
These papers were not published, and
no protective measures were taken.

Some years ago, the corporation of
Aussig, Bohemia, saved a characteristic
rock of basalt, by closing a quarry.
In the year 1906, the corporation of
Vienna granted a sum of about two

million pounds for purchasing 10,872
acres of primeval woods and meadows
around the town.

Moreover, a great many natural
monuments are protected by private
owners. Prince Liechtenstein made
a reservation of 353 acres in Moravia-
Silesia, in which there occur *Juniperus nana* and *Salix herbacea*. Mr
Rothschild, of Vienna, made a reserve
of 1,136 acres of primeval forest
in the mountains of Dürnstein; and
Prince Schwarzenberg reserved, in the
Böhmerwald, 284 acres for the purpose
of safe-guarding a primitive forest in
central Europe. Indeed, this wood, containing much *Picea excelsa* and *Abies
pectinata*, is of the greatest interest to
naturalists and artists.

## Belgium

On his 65th birthday, King Leopold II handed over to the nation several estates, some of great natural beauty, which were menaced by being built over, on condition that the forests which they contained should not be cut down, and that the other parts should be preserved unchanged.

The Belgian government, in 1894, instructed the superintendents of the waters and forests to see that the scenery should not be altered by works of cultivation. Trees, remarkable for their shape or size, or from historic events associated with them, were to be well protected.

Moreover, there exists a "Société nationale pour la protection des sites" which has obtained no little influence.

## The British Empire

In the United Kingdom and its possessions, precedents already exist in the way of registering, mapping, protecting, and publishing accounts of natural monuments.

### *The United Kingdom of Great Britain and Ireland*

#### *Registration*

The whole of England, Wales, and Ireland has now been geologically surveyed. The first maps were published in 1834, but it was not until the following year that the Geological Survey was established as a separate institution. Most parts of Scotland have also been surveyed by the Geological Survey. On such maps, many natural monuments are marked. Floristic

mapping was commenced in Great
Britain over 60 years ago. In 1843,
H. C. Watson began to represent the
areas of the distribution of species and
genera of plants on separate small
charts, all on the same scale[1].

A Vegetation Survey of the British
Isles was commenced by the late Mr
Robert Smith, who published his first
maps in 1900. Mr Robert Smith was
a pupil of that brilliant plant geographer
Professor Charles Flahault, of Mont-
pellier. Flahault's method of repre-
senting vegetation by recording the
distribution of dominant forest trees on
a map[2] was at first proposed in Scotland;
but in view of the general absence
of primitive woodland and the great
proportion of cultivated land in Britain,

[1] Watson, H. C. The Geographical Distribution
of British Plants. 3rd ed., London, 1843.

[2] Flahault, Charles. Essai d'une Carte Botani-
que et Forestière de la France. *Ann. de Geog.*,
No. 28, 1897, pp. 289–312 (with map).

R. Smith decided that this method would give an inadequate if not an erroneous impression of the present vegetation of that country[1].

In 1904, the Central Committee for the Survey and Study of British Vegetation was founded by C. E. Moss, W. G. Smith, A. G. Tansley, and T. W. Woodhead. The functions of this Committee are to bring into close contact all British students who are actively investigating vegetation, to proceed with the mapping of the plant associations of the British Isles, to secure uniformity of method, and generally to promote the survey and study of British vegetation. The Committee has issued a leaflet of suggestions as to methods of survey[2],

[1] Smith, R. On the Study of Plant Associations. *Natural Science*, XIV. London, 1899, p. 109.

[2] Suggestions for Beginning Survey Work on Vegetation. [Out of Print.] See *The New Phytologist* [London]. Cambridge, 1905 (pp. 97–102).

scale of maps (1 : 63,360), colours, and symbols to be used on vegetation maps of this scale. It is actively engaged in furthering in every way the study of vegetation and the publication of results. The Secretary is Dr W. G. Smith, of the Agricultural College, Edinburgh. The following memoirs, which include fully coloured vegetation maps, have already been issued:

*England.*—"Yorkshire : Leeds and Halifax District," by Drs W. G. Smith and C. E. Moss, *The Geographical Journal*, 1903. "Yorkshire : Harrogate and Skipton District," by Dr W. G. Smith and Mr W. M. Rankin, *The Geographical Journal*, 1903. "Basins of the Rivers Eden, Tees, Wear and Tyne," by Mr F. J. Lewis, *The Geographical Journal*, 1904. "Somerset : Bath and Bridgwater District," by Dr C. E. Moss, *Royal Geographical Society*, 1907.

*Scotland.*—"Edinburgh District," by Mr R. Smith, *Scottish Geographical Magazine*, 1900. "Northern Perthshire District," by Mr R. Smith, *Scottish Geographical Magazine*, 1900. "Forfar and Fife District," by Dr W. G. Smith, *Scottish Geographical Magazine*, 1904 and 1905.

*Ireland.*—" South Dublin District," by Dr G. H. Pethybridge and Mr R. Ll. Praeger, *Proceedings of the Royal Irish Academy*, 1905.

The work now in progress includes the following surveys: the Peak District, by C. E. Moss; Cambridgeshire, by C. E. Moss; Hampshire and North-west Yorkshire, by W. M. Rankin; North-east Yorkshire, by W. G. Smith; West Kent, by A. G. Tansley; the Chiltern Hills, by A. G. Tansley; North Dublin District, by G. H. Pethybridge and R. Lloyd Praeger; and Glendalough District, Co. Wicklow, by R. Ll. Praeger.

I cannot but remark that these carefully executed and fully coloured vegetation maps of various parts of the British Isles, showing the distribution of different types of vegetation over the country, and accompanied by descriptive memoirs, represent a standard of voluntary work which has not

been attained in any other country.
One of those maps (South Dublin)
has been printed by the Ordnance
Survey Department for the Irish
Board of Agriculture, and the other
maps have been published either by
the Royal Geographical Society, or
by the Geographical Society of
Edinburgh; and the broad-minded-
ness of these learned societies in this
regard has placed plant geographers
of all nations under a debt of gratitude.
It is, however, satisfactory to know
that the Irish experiment is being
followed with regard to the publication
of vegetation maps of Great Britain;
and that two maps, of the Peak
District, uniform in area and size
with the new series of Ordnance
maps, are to be printed and issued
by the Ordnance Survey Department
immediately, and accompanied by an
explanatory memoir.

Dr M. Hardy has published a general account of the vegetation of the Highlands of Scotland, accompanied by a map on a smaller scale[1].

An elaborate study, including the detailed mapping out, of a salt marsh in Brittany is being carried out under the direction of Professor F. W. Oliver, of the University of London. Several annual reports dealing with the progress of the work and the methods employed have already appeared[2].

Mr F. J. Lewis is also engaged[3] on a study of the succession of deposits in the peat of Scotland; Dr T. W. Woodhead is continuing[4] his investi-

[1] M. Hardy. Des Highlands D'Écosse. Paris, 1905.

[2] Cf. *The New Phytologist*, 1904 to 1908.

[3] Lewis, F. J. The Plant Remains of the Scottish Peat Mosses. *Trans. Roy. Soc. Edin.*, xli. part iii., No. 28, 1905; xlv. part ii., No. 13, 1906; xlvi. part i., No. 2, 1907.

[4] Woodhead, T. W. Ecology of Woodland Plants in the neighbourhood of Huddersfield. *Linn. Soc. Journ., Bot.*, 1906.

gations of the woodlands of West Yorkshire; Professor F. E. Weiss is studying and mapping the vegetation of Delamere Forest, Cheshire; and Professor R. H. Yapp is continuing[1] his experimental studies on the ecology of Wicken Fen, Cambridgeshire.

Pictorial records or photographs of natural monuments are much to be desired, and the British Association Geological Photographs' Committee, the British Association Botanical Photographs' Committee are already at work on these lines. Besides this, the Central Committee for the Survey and Study of British Vegetation collects photographic prints showing features of ecological interest and illustrating definite members of plant associations. These collections, at present, are lodged at University College, London, and duplicate collections are deposited at

[1] Yapp, R. H. Wicken Fen. *The New Phytologist*, 1908.

the Botany School of the University of Cambridge, and at the University of Manchester.

Of course, many other illustrations of natural monuments, besides photographs, are in existence; and it is not necessary to mention all. I cannot, however, help citing Gowan's Nature Books, which contain a great number of photographs of plants and animals in their native haunts.

### Protection

The protection of natural monuments in Great Britain is not so easily carried out as in continental States. Still, something has already been done by government, public corporations, and other bodies; and hence not a few British natural monuments are preserved.

GOVERNMENT.    According to a
report of the Office of Woods, Forests,
and Land Revenues, about fourteen
State Forest Reservations exist.

For instance, the Forest of Dean
in the western part of the county of
Gloucester comprises 24,000 acres, of
which the wooded area occupies about
18,500 acres.   The crown is entitled
to keep enclosed an area of 11,000
acres against all rights of common
over the surface;   but, in 1896, the
area actually enclosed was not more
than 4,600 acres, the remaining 13,900
being open woodland.   During the
greater part of the eighteenth century,
the forest was managed with a view to
growing oak for the navy.   With the
introduction of ironclads, the demand
for navy timber ceased; and the crown
was left with a crop that was immature,
a good deal of which had been injured
by the throwing open of the enclosures

and the consequent destruction of the underwood. The system of management which is now being pursued on a systematic plan is gradually to re-enclose up to 11,000 acres, with a view to the establishment of a complete crop of mixed beech and oak in high forest, with scattered larch, chestnut, sycamore, and other introduced trees.

The Highmeadow Woods, adjoining the Forest of Dean, are the absolute property of the crown, free from any rights of common, and contain about 3,285 acres of enclosed woods. Two-thirds consist of old timber, and the remaining one-third of trees from 47 to 72 years old. The best part of the woods appears as high forest of old and middle-aged oaks over a coppiced underwood; whilst elsewhere the standards have been more freely cut, and the coppice is vigorous and well intermixed with seedlings chiefly of ash and oak. The system of management now being followed aims at the gradual conversion of 2,286 acres into high forest, and the regular treatment of 999 acres as coppice and standards.

Windsor Forest contains two parks, the Great and the Home Park. Windsor Castle itself is nearly surrounded by the latter, which, with the exception of about 70 acres open to the public, forms the private pleasure grounds of the castle. The Great Park forms a portion of the Royal Domain attached to the castle, but the public are allowed access to the greater part of it. In addition to the parks, there are also belonging to the crown large areas of woodland in the vicinity (formerly included in Windsor Forest, now known as the Oak Forest, and adjoining the Great Park), and the woods at Swinley, Easthampstead, Bagshot, and Ascot. The Great Park and the wood-

lands together contain in area upwards
of 14,000 acres, which consist either of
woodland, or mixed heath and woodland, or pasture in the park.

The New Forest, Hampshire, has
an extent of about 92,365 acres, of
which some 30,000 acres are private
property, and the remainder is a State
Forest Reserve proper. The existence
of certain rights of pasturage has prevented the enclosure of the New Forest,
except to a limited extent. The State
Forest Reserve consists approximately
of 40,000 acres of open heath and pasture without timber, 5,000 acres of heath
and pasture with timber, 11,250 acres
of enclosed woodland and 6,400 acres
of unenclosed woodland. The Forest
contains primeval woods, moor, heaths,
and small peat mosses. There are
some streams but no lakes: some small
ponds are however left where the peat
has been removed.

The Esher Woods, in the county of Surrey, are principally self-sown oak wood, and therefore varying greatly in age. The whole area comprises about 844 acres. Of course, the trees of the State Forests of England are often planted; but certain parts possess indigenous woodland and heath. Many animals, particularly the smaller ones, are protected by the Forest Reserves.

PUBLIC CORPORATIONS. County councils, town corporations, and other public bodies often possess commons and parks, which remain free and open for ever[1]. These stretches of land occasionally possess indigenous plant associations; and in this way natural monuments may be indirectly protected. Moreover, county councils have often reserved other open places of natural beauty for the public benefit.

[1] Slater, G. The English Peasantry and the Enclosure of Common Fields. London, 1907.

For example, Queen's Wood is maintained by the Hornsey Borough Council. The London County Council maintains 96 open spaces, to the extent of about 5,000[1] acres, for the use and enjoyment of the public. It is true that most of these open spaces are gardens, artificial grass lands, or plantations; but, on the other hand, there are also woods and heaths, partly showing natural vegetation, as the following examples will show.

Bostall Heath and Woods consist of about 133 acres, and possess an undergrowth of ferns, gorse, briers, and brambles, crowning a hill overlooking the Thames Valley. The woods are formed of densely planted Scots pines, which are unfortunately fast dying out. Golder's Hill adjoins the West-heath,

[1] Sexby, F. F. London Parks and Open Spaces under the control of the County Council. London, 1906.

Hampstead. The sloping banks of
one valley in particular are covered
with an indigenous vegetation. Hack-
ney Marsh, of 339 acres, extends to
the eastern boundary of the London
county; and it is partly skirted and
partly intersected by the original course
of the river Lea. Hainault Forest,
in the county of Essex, is the largest
single area under the control of the
council. It comprises about 805 acres,
and is part of the area once covered by
the historic forest of Hainault, which,
like Epping Forest (page 71), was
anciently included in the great forest
of Essex or Waltham, and was a
favourite royal hunting-ground. Until 1851, the forest of Hainault retained
its original woodland character; but
under an Act passed in that year, the
tract was disafforested, the timber of
the greater part of the forest grubbed
up, and the area converted into arable

land. Nevertheless, some spontaneous plant associations and wild animals are to be found there at the present time.

Hampstead Heath, of an undulating character, has a magnificent position on the heights on the north-western edge of the London county. It has an extent of 2,405 acres, and is well supplied with various ponds. Here and there trees and shrubs are to be seen, such as oak, Scots pine, hornbeam, elm, service tree (*Pyrus torminalis*), hazel, and furze. At various points, enclosures and plantations have been formed as sanctuaries for bird-life.

Marble Hill, the centre of the famous view from Richmond Hill, was, in 1901, put up for sale as building land. The trees were being felled, and there was imminent danger of the beauty of the view being irrevocably destroyed. The charm of this landscape has been the joy of many people, and its delights have been sung by poets and immortalized by painters. The sweeping curve of the river forms one of the most beautiful of Nature's pictures in all Britain. By means of gifts from private donors and various public authorities, £36,663 were raised, and the remainder was paid by the London County Council. Thus a place of great natural beauty has been saved, partly by private generosity, for the public use and enjoyment.

Streatham Common, of about 66 acres, is situated at the southern extremity of the county. The upper part is undulating, and covered with clumps of trees, gorse, brambles, and other undergrowth; whilst the lower part is more open and available for games.

The Corporation of London has perhaps done more to preserve natural monuments than any other body in the world. Several landscapes have been bought in the environment of London for the protection of their original aspect as far as possible.

During the years 1877 to 1884, Epping Forest, of 5,610 acres, was bought for £211,000. This includes the land, the timber, and the abolishment of the rights of lopping. It lies upon a moraine, and possesses natural woodland, heaths, and ponds. Its indigenous plants include oak, beech, hornbeam, birch, holly, apple, gorse, and willows. Besides this, very little planting has taken place for many years; and it is, generally speaking, a natural forest which has been much disfigured by the custom of lopping in former times. It is the duty of the committee to preserve at all times, as far as possible, the natural aspect of the forest, and to protect the timber. The only cutting-down allowed is the necessary thinning-out for the protection and growth of the timber. Hunting in the forest is not exactly allowed, but is tolerated in the northern part. Fishing in the ponds is also permitted.

In the year 1883, Burnham Beeches, near Slough (figure 5), were bought for £7,624. They form a beautiful stretch of woodland, 375 acres in extent, growing upon a moraine. In this natural forest, the beech is prominent, but oak, hornbeam, birch, hazel, holly, spurge laurel, willows, Scots pine, and juniper also occur. The beeches are in part very much deformed by having been lopped in former times. They are ten to twenty feet round, and I cannot remember having seen bigger ones anywhere. The trunks are for the most part hollow, and in some cases adventitious roots originate in the interior and go down into the soil. The oaks, too, attain a circumference of fifteen feet, and their old trunks are hollow. There

Figure 5. Burnham Beeches, preserved by the Corporation of London.

is also a little pond with white and yellow water-lilies. Burnham Beeches are carefully protected, and here and there old trees are supported by wooden props and iron rings.

In 1886, Highgate Wood, 69 acres in extent, was given over to the Corporation by the Ecclesiastical Commissioners for preservation. The wood is indigenous, and consists chiefly of oak and hornbeam, with also holly and wild cherry.

Following the example of London itself, several provincial towns, for example, Richmond, and also some Scottish cities, have made reservations. One example may be mentioned here. In Victoria Park, Glasgow, a so-called fossil grove of the Carboniferous formation has been preserved (figure 6). The stumps of several fossil trees, which appear to be those of Lepidodendron, varying in diameter from

about one to three feet, are fixed
by long forking "roots" in a bed of
shale. Here and there, the spread-
ing "roots," which bear the surface
features of Stigmaria, extend for a
distance of more than ten feet. In
one place, a flattened Lepidodendron
stem, about thirty feet long, lies prone
on the shale. Near one of the trees
and at a somewhat higher level than
its base, the surface of the rock is
clearly ripple-marked, and takes us
back to the time when the sinking
forest trees were washed by waves
which left an impress in the soft mud
laid down over the submerged area[1].
Over this relic of ancient times, a hall
has been erected, for protecting these
interesting objects against the influence
of the weather. If I am right, this

---

[1] Seward, A. C. Fossil Plants for students of
botany and geology. With illustrations, Vol. I.,
Cambridge, 1898, p. 57.

Figure 6.  Carboniferous Forest, Victoria Park, Glasgow.

Fossil Forest Reservation is the only one of its kind in all Europe; but I was recently informed that it is not looked after carefully, as a good many plants of recent growth have sprung up which have no place in a fossil grove. If so, it is to be wished that the original state of this monument of nature may soon be restored.

SOCIETIES. There are many associations, societies, and clubs whose object it is to promote the preservation of beautiful scenery and natural objects of great interest. Perhaps the most important of its kind is the National Trust for Places of Historic Interest or Natural Beauty, formed in 1895, and incorporated under the Joint Stock Companies Act. The Trust is the only corporation in Great Britain, with the exception of County Councils and the Board of Works empowered to hold lands or buildings of beauty

Figure 7. Natural Monuments preserved by the
National Trust.

1. *Barmouth.* The beautiful cliff known as "Dinas-o-leu," overlooking the estuary at Barmouth. It is four and a half acres in extent.

2. *Barras Head.* Fifteen acres of cliff land at Tintagel, looking on to the magnificent pile of rocks on which stand the ruins of King Arthur's Castle.

3. *Toy's Hill.* Heathland on a Kentish hill-side, overlooking the Weald.

4. *Ide Hill* near Toy's Hill. Fifteen acres of wooded hill-side.

5. *Wicken Fen.* Strips of land, together containing about four acres. Wicken Fen is almost the last remnant of primeval fenland in East Anglia.

6. *Derwentwater.* Nearly 108 acres of the western shore of Derwentwater are already protected. Negotiations are in progress for the acquisition of additional land of great charm. An option is also held of buying land at the south end of the lake. A waterside walk round the lake is also contemplated.

7. *Kymin.* Nine acres on the summit of the Kymin Hill at Monmouth, which commands magnificent views of the valley of the Wye and the Monnow.

8. *Rockbeare.* Twenty-one acres on the top of Rockbeare Hill, Devonshire, covered by trees and heather.

9. *Hindhead.* Seven hundred and fifty acres of Common land on the summit of Hindhead, Surrey. The preservation of a neighbouring tract of land, including 17½ acres of woodland, is also contemplated. Very recently, the Trust has obtained a valuable tract of heathland and a beautiful hillside, together occupying about 65 acres.

10. *Newtown Common.* Near Newbury.

11. *Ullswater.* Gowbarrow Fell and Aira Force, 750 acres in all, bordering on the lake.

12. *Burwell Fen.* Thirty acres of Fenland.

13. *The Grey Wethers.* Two plots of land, together about twenty acres, in Piggle Dene and Lockeridge Dene, near Marlborough, on which are large and characteristic examples of the Sarsen stones known as the Grey Wethers.

or historic interest, for the public good. The board of the Trust is partly elected by the members, partly by the chief learned bodies and societies in the country, and it is thought that a body thus governed provides the most efficient organisation for the suitable management of property requiring peculiar treatment. Hitherto not only ancient buildings and historic monuments, but other properties of various kinds, such as hills, cliffs, fens, and woods, have been acquired by the Trust (figure 7).

In 1895, a beautiful bit of Welsh cliff at Barmouth, of four to five acres in extent, was presented to the Trust by Mrs Fanny Talbot. It is to be kept, as far as possible, in its natural condition, and particularly the furze is to be unmolested. In the year 1897, Barras Head, on the north Cornish coast, was purchased for

£505, from the late Earl of Wharn-
cliffe, the money being raised by
means of donations. The fifteen acres
of cliff land which form the top of this
promontory, and the original vegetation, will thus be preserved inviolate.

Nearer London, too, in a district
which is already being invaded by
villas, the Trust has secured two hills
overlooking the Weald. Toy's Hill
was given to the Trust in 1898 in
memory of the late Mr Frederick
Feeney, by Mr and Mrs Richardson
Evans; and in the following year,
Ide Hill, fifteen acres of wooded hill-
side, was purchased for £1,636, by
means of donations.

In the same year, 1899, two small
strips of Wicken Fen, Cambridgeshire, were acquired. This Fen is
nearly the last remnant of the original
fenland of East Anglia, and is of
special interest to botanists and ento-

mologists, on account of the occurrence there of rare plants and insects, and of primitive fenland plant associations. The two strips of land, about four acres in all, were obtained in 1899. The first was purchased from Mr J. C. Moberley of Southampton, and the second given to the Trust by the Hon. N. Charles Rothschild, of London. It is hoped that the whole of Wicken Fen, which is only of comparatively small extent, will eventually be acquired.

In 1902, the Brandlehow Park Estate on the western shore of Derwentwater, was purchased for £6,500 raised by means of public subscriptions. Hitherto the shores of Derwentwater had been open to the public on sufferance only; but now a large area is made accessible both by land and from the lake. It is a varied tract of meadow and woodland, about 108 acres in extent, where the public may walk without restraint, and enjoy romantic views of typical lake scenery.

In the same year, also, nine acres on the summit of the Kymin Hill, Monmouth, were purchased for £300, by means of donations. The spot is about 800 feet above sea level; and it is well known as affording one of the finest prospects in the west of England, for it commands an outlook extending over nine counties.

In the year 1904, Rockbeare Hill, Devonshire, a few miles east of Exeter, was presented to the Trust by Mr W. H. C. Nation, of Rockbeare. The land, which is 21 acres in extent, is characteristic of the south-west of England, and is partly covered by a wood, partly by open heath.

Further, in 1906, the manorial rights over 750 acres of common on the summit of Hindhead were acquired. This land was offered for sale by public

auction ; and it was felt that, in a popu-
lous and growing neighbourhood like
that of Hindhead, common land ran
great risk of disfigurement and en-
croachment while in private hands.
The Hindhead Preservation Committee was therefore formed, at the suggestion of the Commons' Preservation
Society (see below), and a few public-
spirited persons raised a guarantee,
on the strength of which the land was
bought for the sum of £3,625. The
Committee then presented the land to
the Trust. This land also includes
the Devil's Punch Bowl, a deep
combe of the most striking character,
surrounded by open heathland, rising
to nearly 900 feet above the sea,
covered with heather and furze, and
dotted over in the more sheltered places
with birches, hollies, and Scots pines.
The preservation of a neighbouring
tract of land, including $17\frac{1}{2}$ acres of

woodland, is also contemplated. Quite recently, the Trust has obtained a valuable tract of heathland and a beautiful hillside, together occupying about 65 acres.

In the same year, Newton Common, near Newbury, a tract of land bearing beautiful trees and commanding extensive views over the surrounding country, was also saved. Mr W. T. Shaw, of Horris Bank, who was anxious to preserve the wood and prevent the erection upon the land of any unsightly building injurious to the beauty of the Common, bought the land and presented it to the Trust.

In 1906, too, Gowbarrow Fell and Aira Force, 750 acres in all, bordering on Ullswater, were purchased for £12,800. It is a beautiful tract of woodland, lake, and fell, with the exquisite glen through which the Aira tumbles and leaps to the lake below. The very large sum raised by public subscription had been got together in certain large northern towns, chiefly Edinburgh, Manchester, Liverpool, and Newcastle.

Last year, thirty acres of Burwell Fen, adjoining Wicken Fen, were presented to the National Trust by Mr Charles Rothschild. This fenland offers a much-needed sanctuary to rare plants and insects of that district; but, having been once reclaimed, it is not nearly so valuable, from this point of view, as Wicken Fen.

Only this year, the Trust has acquired two plots of land, together consisting of about twenty acres, in Piggle Dene and Lockeridge Dene, near Marlborough, on which are characteristic examples of the Sarsen Stones known as the Grey Wethers.

Scientific societies have also given proof of their interest in the preservation of natural monuments. Thus, the Council of the Geological Society of London has just contributed, from the Barlow Jameson Fund, a sum to secure the preservation of the above mentioned Sarsen Stones, called the Grey Wethers (page 10). The Selborne Society, founded in London in 1885, aims to promote the study of natural history, to preserve wild plants and animals from needless destruction, and to protect places of natural beauty and of antiquarian interest. The Society arranges lectures and publishes a monthly journal, *Nature Notes*, which contains papers on the preservation of natural monuments. The Entomological Society of London also has a committee for the protection of insects. But beyond issuing a list of those Lepidoptera, which the committee thought ought to be protected, it has done nothing as yet in the matter.

The societies for the protection of birds and other similar associations co-operate for the preservation of some kinds of natural monuments. The Royal Society for the Protection of the Birds of London offered, in 1908, the gold medal of the Society and a cash prize of 20 guineas for the best essay or treatise on comparative legislation for the protection of birds.

An instance of the direct result of the efforts of the Irish Society for the Protection of Birds may be given here. At Malahide Island, there is a nesting-place of the Common and Arctic Terns. A few years ago a couple of pairs of the Common Tern were nesting ; and latterly both species have bred there, and on one of the islands at Skerries. The society in question has appointed a watcher, whose duty it is to see that visitors do not molest the birds or take their eggs.

In consequence of this, last year about two hundred nests of both species of Terns and more than three hundred eggs were observed[1]. It is to be wished that the success attending the efforts of the Irish Society for the Protection of Birds may be imitated by other societies in other parts of the British Isles.

Again, associations and clubs of public benefit have co-operated for the protection of natural monuments. The Kyrle Society, founded in 1877 aims at bringing beauty, in the widest sense of the term, home to the people. The society undertakes to secure open places and to prevent spaces being illegally built upon, and to co-operate with local societies for the preservation of commons and places

[1] Williams, A. Wild Bird Protection in Co. Dublin. *The Irish Naturalist*, Vol. XVII. Dublin, 1908, p. 119.

of natural beauty.   There are also branches of the Kyrle Society in the provinces, for instance, in Bristol, Cheltenham, Dublin, Glasgow, Liverpool, and Nottingham.

The Commons' and Footpaths' Preservation Society, founded in 1865, aims at preserving not only commons, but also village greens and waste lands ; and it is anxious to assist local authorities to secure open spaces by purchase or otherwise, and to promote arrangements with landowners for the opening of places of natural beauty for the use of the public.   Amongst other things, the society rendered considerable assistance to the movement which resulted in the preservation of the extensive fine view from Richmond Hill.   Further, the society helped to preserve the East Cliff, near Hastings, threatened by a Railway Company with serious disfigurement.

The National Society for Checking the Abuses of Public Advertising has for its object the retention of the picturesque simplicity of rural and river scenery. It attaches much importance to teaching the young to find pleasure in nature and in the picturesque aspect of every-day scenes, and impresses on all that they ought to respect the right of others to the unimpaired enjoyment of the country and open spaces. The organ of the society, called " Beautiful World," also contains articles on the care of natural beauty.

The Coal Smoke Abatement Society also has the preservation of natural beauty among its objects.

Moreover, there are many local societies which have as one of their objects to preserve, either directly or indirectly, natural monuments of some kind or other. The following are some such societies :—the Thames

Preservation League; the Society for
Preserving the Natural Beauty of
Criccieth, Port Madoc, Harlech and
neighbourhood; the Society for the
Preservation of the Beauty of Hythe
and neighbourhood; the John Evelyn
Club for Wimbledon; the Beautiful
Oldham Society; the Hampstead
Heath Preservation Society; the
United Devon Association and, con-
nected with it, the Western Counties
Ferns' and Wild Flowers' Preservation
Committee; and a great many Natu-
ralists' and Field Clubs.

For example, the Yorkshire Naturalists' Union, with over 4,000 members,
and over 40 affiliated societies, en-
courages the preservation of natural
monuments in many ways. The
Union's official organ, *The Naturalist*,
also advocates the preservation of
plant and animal life, in their natural
condition as far as possible. The Wild

Birds' Committee has shown practical interest in this matter by paying for a watcher on the Spurn peninsula during the whole of the breeding season. In the same way, birds on the Flamborough headland and on the western fells are protected. The Glacial Committee also has taken steps to preserve the larger and more important ice-carried boulders, by arranging for their removal to public parks or other suitable sites, whenever it was impossible to preserve them *in situ*. The Geological Photographs' Committee record, by means of photographs, many interesting geological sections as they become exposed ; and this is often the only practicable method of preserving them, as they are often destroyed shortly after exposure, sometimes by quarry-owners and sometimes by the action of the sea. In addition, the Union, by the publication of its various

monographs and transactions, is taking
a full record of the Geology, the
Flora, and the Fauna of the whole
of the county of Yorkshire.   In these
monographs, special care is taken to
give all available particulars of disappearing plants or animals.

Finally it must be mentioned that,
two years ago, an English branch of
the League for the Preservation of
Swiss scenery was constituted.   It
has for its object to save, as far as
possible, the natural and historic charm
of Switzerland from the harmful influ-
ences of commercial enterprise.   It
seeks to check the disfigurement of
scenery, and to prevent the undue
multiplication of railways, the erection
of incongruous buildings and advertisements, all of which have had such
a rapid extension of recent years, and
which are at the same time detrimental
to national life.   Members, when visit-

ing Switzerland, are asked to impress upon hotel keepers and others the truth that unimpaired natural beauty is their chief asset, and that, whilst they cannot add to the attractions of their country, to diminish them is a short-sighted and a fatal policy.

INDIVIDUALS. Great Britain is fairly fortunate, as the great landowners manage their estates on conservative lines. In many cases, woods and moors and even small lakes are preserved for the sake of their beauty, or for the plant and animal life which they sustain. Some of the Norfolk broads are so preserved. Moreover, British landlords have a passion for carefully protected game-preserves. It would be almost impossible to enumerate here all examples of such a nature, but one case from Ireland may be mentioned. The island of Lambay, near Dublin, is a large breeding-place of sea birds, and

also a locality of much botanical interest. It is now carefully protected by
the owner, the Hon. Cecil Baring.

On the other hand, there are always
persons who are ready to subscribe
money for preserving a place of scientific interest or natural beauty, as the
reports of the above mentioned societies
show.

### British Colonies and Dependencies

In the British possessions, much is
being done in the way of protecting
original nature.

*East Africa.* Shooting is strictly
prohibited over a considerable stretch
of territory along the Uganda Railway.

*Rhodesia.* At the instigation of
the late Mr Cecil Rhodes, a National
Park has been instituted in the
Matoppo Mountains. For the care

of the native animal world, part of the district is enclosed by a fence.

*Cape Colony.* According to a Proclamation it is not lawful at any time to kill or pursue the following sea birds:— the jackass, penguin (*Spheniscus demersus*), the Cape gannet (*Sula capensis*), and the common and the Cape cormorants (*Graculus carbo, G. capensis*).

*Seychelles.* On the isles of Praslin and Curieuse, all places where the Lodoicea palm grows have become crown lands.

*India.* An act, dating from 1873, prevents the indiscriminate destruction of wild elephants; and another act was passed in 1879 for the protection of these animals. Further, by an act of 1902, the exportation of skins and plumes of all birds except poultry is prohibited. A further exception is made in favour of specimens for educational and scientific purposes.

*Bengal.* In the Darjeeling District, any European is allowed to take an orchid or fern in the forest and to grow it in his own house ; but if a larger quantity is required for his own use, he is given a permit, but no permit is given to collect the plants for trade.

*Assam.* In general, the removal of orchids from the Faintia Hills and the Khasi Hills is prohibited by the Deputy Commissioner ; but passes for six months were granted for collecting some common species up to a distinct limit.

*Ceylon.* The jungle, above 5,000 feet, must not be cut.

*West Australia.* A large part of the primeval forest in Darling Range has been reserved by Parliament. Some of the most remarkable plants are:—*Banksia attenuata, B. grandis, Casuarina Fraseriana, Eucalyptus calophylla* (Red Gum), *E. marginata*

(Jarrah), *Kingia australis, Macrozamia Fraseri, Nuytsia floribunda,* and *Xanthorrhoea Preissi* (Diels).

*Queensland.* In the hinterland of Trinity Bay, there is a reservation on Bellenden Ker Range, of about 86,450 acres. It includes the highest North Australian mountains, covered below with jungle (*Agathis, Araucaria, Calamus, Musa,* Orchids, *Podocarpus amara*), and at the top with characteristic mountain types (*Dracophyllum, Rhododendron Lochae*). This reservation is of very great biological importance, for few districts in the whole of Australia are so crowded as this with endemic species of the vegetable and animal worlds.

*New South Wales.* There is a National Park (Eucalyptus trees) near Sidney, which has an extent of about 37,050 acres. It is to be kept in its original state, and the birds and the animal world generally are strictly protected.

*Victoria.* Ferntree Gully, near Melbourne, a tract with *Alsophila australis, Dicksonia antarctica, Eucalyptus*, and *Eugenia Smithii* growing on it, has become crown land. The taking away of ferns particularly is forbidden.

*New Zealand.* In the first place, a Committee has been formed at Christchurch for purchasing as a Public Reserve, Kapiti Island in Cook Strait. This is an original forest of evergreen trees, and coniferous trees such as *Podocarpus dacrydioides, P. spicata*, and *P. totara.* Another reservation is Little Barrier Island, with *Agathis australis, Metrosideros tomentosa*, and other types of Northern New Zealand[1]. The island also contains a

[1] See also Laing and Blackwell, Plants of New Zealand.

bird sanctuary[1]. Resolution Island, with *Dracophyllum longifolium, Veronica elliptica*, and other rare plants, is also preserved.

## Denmark

The Danish Parliament, induced in part by Professor E. Warming, of the University of Copenhagen, granted a large sum of money for preserving several natural landscapes. In the years 1900 and 1901, Raabjerg Mile was bought for £100. This is situated near the cape of Skagen, and is the largest dune in Denmark, occupying 469 acres. About 47 acres are devoid of vegetation, and it is for the future to remain uncultivated. In the neighbourhood, there are smaller dunes with

[1] Drummond, F. The Little Barrier Bird Sanctuary. *Trans. New Zealand Inst.*, Vol. XL. Wellington, 1908, p. 500 *et seq.*

*Psamma arenaria, Hordeum arena-
rium, Salix repens,* and *Viola tri-
color* ; and also some dune ponds, with
such interesting plants as *Pilularia
globulifera, Sparganium affine, Ranunculus reptans, Subularia aquatica,
Bulliarda aquatica, Radiola linoides,
Elatine hexandra, Centunculus mini-
mus, Littorella lacustris,* and *Lobelia
Dortmanna.*

In the following year, Fossedal was
purchased for £400, on condition that
it should be preserved in its original
state.     Fossedal is a valley of 1,235
acres, with an adjacent dry heath,
containing *Empetrum nigrum, Arctostaphylos Uva-ursi, Calluna vulgaris,*
and other heath-loving species.

Thirdly, Borrishede, a heathy tract
of ground of about 3,705 acres, was
bought for £1,600, subject to rifle-
practice being allowed there during
a few weeks every year.     For the

most part, the heath is dry and covered
with *Aira flexuosa*, *Salix repens*,
*Empetrum nigrum*, *Vaccinium Vitis-
Idaea*, *Arctostaphylos Uva-ursi*, *Cal-
luna vulgaris*, and such lichens as
*Cladonia furcata*, *C. rangiferina*, and
*C. uncinatis*. At other places, there
are heathy moors with *Andromeda
Polifolia*, *Juncus squarrosus*, and
*Erica Tetralix*.

By a law passed in 1894, the white
and the black stork and all singing
and other small birds are protected
the whole year round.

Several natural monuments are
protected by the government. Vrö-
gumhede, a tract of dry heathy ground
of 99 acres near Varde, is excluded
from cultivation of any sort. Further, a
number of large boulders have been
saved as ancient monuments (fredlyst)
because old traditions are connected
with them. For example, Tislunde

Stenen between Esbjerg and Kolding,
1·9 to 3·8 m in diameter and 3·8 m in
height; and Laeborg Stenen near the
church of Laeborg and in the same
county, 3·8 m in length and 1·0 to
1·7 m broad and 0·9 m high above the
soil; Damme Stenen, frequently called
Hesselager Stenen, on the isle of
Fünen, with an approximate height of
12 m and a circumference of about
45 m—the largest boulder in all
Denmark; Rokkestenen in the vicinity
of Nexö, on the isle of Bornholm, a
balancing stone of 2·2 to 4·4 m diameter
and 3 m height, and given over by
the proprietor to the custody of the
Mineralogical Museum, Copenhagen
University : all these are protected
and conserved by the government. A
pecuniary penalty is imposed on any
one who damages the last-named
natural monument.

Societies and private individuals

have also promoted the care of rare plants and animals. By the proposal of the Botanical Society of Copenhagen, in the University Forest near Alindelille, *Ophrys muscifera* and other orchids are protected. The Danish Ladies' Association for the protection of animals is also anxious to form sanctuaries (frednings-steder). Near Aalborg, one such sanctuary of 68 acres extent has been instituted, for preserving not only its birds, but its animals and plants.

In the spring of 1905, I was invited to lecture before the Natural History, the Geological, and the Botanical Societies of Copenhagen, on the care of natural monuments, particularly in Denmark. The minister of the education department and the professors of the university and academies were present. In the same year, a committee for the protection of nature

(Udvalg for Naturfredning) was formed by the three above mentioned societies. At the suggestion of this committee, a wood affected by the prevailing winds near Tisvilde on the isle of Zealand, and the weeping beeches in two forests have been protected. In Ringköbing Fjord, a part of a virgin island, the whole animal world has been saved. Lastly, a colony of herons' nests at Jonstrup Vang near Copenhagen has been preserved.

The late Th. Schiötz, of Odense, bought a piece of heathland near Ringköbing, the only Danish habitat of *Arctostaphylos alpina*, and gave it to the Danish Heath Association (Dansk Hedeselskab), for the preservation of the plant. He protected, in a wood near Buderupholm, the only Danish habitat of the Lady's slipper (*Cypripedium Calceolus*) by enclosing it with a hedge. H. Winge, keeper of the

Zoological Museum Copenhagen, reserved a small coast-district, for preserving its characteristic vegetation
and animal life.

## France

Efforts for preserving natural monuments have been promoted for some-
time past by the State Forest Board ;
and it is worthy of note that the
higher officials of the forests, corre-
sponding to the German "Oberforst-
meister," are in France called "Con-
servateurs des Fôrets." There is an
order, dating from 1827, that in the
State Forests no ground-growths of
any kind are to be removed and
utilised without strict permission of
the Board.

By a law of 1861, several portions
of the State Forest of Fontainebleau
(les séries artistiques) have been ex-

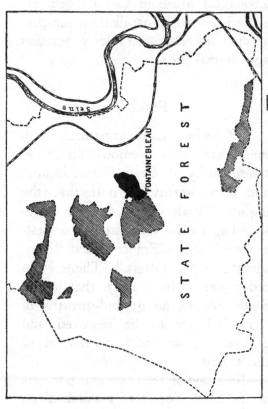

Figure 8. Chart of French State Forest, showing portions ▨ protected by law.

cluded from the course usually followed;
and they are to be preserved unchanged for artists and for the enjoyment of the Parisians (figure 8). In
1899, it was ordered that a register of
all remarkable trees should be drawn
up. The trees were to be protected;
and, when this was impossible, a special
report to the Board was to be made,
and the forest inspectors were charged
with the inspection of these trees.

In the year 1900, the prefects of
the provinces of Isère and Savoie prohibited the uprooting and the exposing
for sale of alpine plants, such as *Cypripedium Calceolus*, *Eryngium alpinum*,
*Gnaphalium Leontopodium* (Edelweiss), *Rhododendron* spp., *Cyclamen
europaeum*, *Artemisia Mutellina*.

In 1906, a law was passed organising the protection of landscapes and
natural monuments of an artistic but
not of a scientific interest. Besides this,

in each province a committee is to be
constituted, and the Prefect is to be in
the chair. The committee has to make
out a list of places of natural beauty,
the protection of which might be of
interest from the artistic point of view.
The proprietors of such localities are
to be invited to bind themselves
neither to destroy them nor to change
their character. If this is agreed
upon, the natural monument is to be
entered on the list, by order of the
Minister; but if not, the Prefect or
the Mayor may be authorised to institute proceedings of expropriation.
Once such a change has been made,
any alteration is punishable with a
penalty of 100 to 3,000 francs.

A "Société pour la protection des
paysages" was founded in 1902. This
society publishes a quarterly bulletin,
which contains many papers on the
care of natural monuments.

## Germany

The Imperial Post Office has forbidden any one to cut the branches of, or otherwise to injure, any tree for the sake of the telegraph wires passing along the roads.

### *Bavaria*

So far back as a hundred years ago and more, a case can be quoted where natural beauty was protected by the government.   In 1803, a wood near Bamberg was bought up by the State for preserving this lovely place for the benefit of the citizens of that old town[1].

In 1840, the government issued an order for the protection of glacial boulders and other remarkable rocks.

[1] Conwentz, H.   Schutz der natürlichen Landschaft, vornehmlich in Bayern.   Berlin, 1907, S. 1.

In 1842, it was decided that avenues of trees were not to be violated or cut down without authorisation by the King.    In 1846, King Ludwig I preserved by purchase a beautiful old oak near Moosach.    In 1852, an edict was issued for the protection of old lime-trees, oaks, beeches, elms, and other trees standing in villages and country towns.    In the following year, induced by the Bavarian Academy of Sciences, the government ordered that glacial boulders should be registered and marked on the forest maps.    Two years later, the foresters were obliged to preserve all interesting natural features, and to notify any damage caused to them; and in 1884, the making of an inventory of these features of the State Forests was ordered.

As the efforts to protect natural monuments in Prussia had been supported by government, I received an

invitation to lecture on this subject
in Bavaria; and in 1903, at a joint
meeting of the Geographical and
Natural History Societies of Munich,
I gave a lecture on the preservation
of natural monuments, with special
reference to Bavaria. Members of
the Court and delegates of the government attended the meeting.

Then, too, the government was
also induced by the " Deutsche und
Oesterreichische Alpenverein" and by
natural history societies to support the
efforts for the protection of natural
monuments according to a regular
plan. Under the lead of the Home
Department, in 1905, a general committee for the care of nature (Natur-
pflege) in Bavaria was formed by
delegates of the authorities and of
scientific and artistic societies. Local
committees were also established in
various parts of the country, for

instance in Amberg, Ansbach, Aschaffenburg, Augsburg, Bamberg, Hof, Kempten, Landshut, Nürnberg, Passau, Regensburg, Wunsiedel, and Würzburg.

In 1905, the department of forestry charged the forest officers to formulate special records of natural monuments. Some forest officers were engaged to protect rare alpine plants, and it was ordered that the shooting of rare birds or those with beautiful plumage should be restricted.

Several communities have supported these efforts. Some years ago, at Munich, Nürnberg, and Regensburg, bye-laws existed by which it was prohibited to offer for sale wild plants with their roots or bulbs. It would be a matter of satisfaction if similar regulations could be enacted elsewhere for protecting, as much as possible, indigenous vegetation in the neighbourhood of large towns. Further, Nürnberg prohibited the gathering of a rare water-lily (*Nymphaea alba* var. *semiaperta*) growing in a town-pond. Wunsiedel saved the habitat of the little luminous moss, *Schistostega osmundacea.*

The Landshut Natural History Society has preserved the remainder of a heath formation, by actual purchase. The Royal Botanical Society of Regensburg bought a tract of land with *Draba aizoides* and other rare plants of the Fränkische Jura in the Goldberg near Pettendorf. Again, a gypseous hill near Windheim, with a characteristic pontic flora, was purchased by the Nürnberg Botanical Society. This vegetation possesses the following species :—*Stipa pennata, S. capillata, Phleum Boehmeri, Koeleria cristata, Poa bulbosa* var. *badensis, Festuca sulcata, Carex montana, C. hu-*

*milis, Silene Otites, Arenaria serpylli-*
*folia, Thalictrum minus, Anemone Pul-*
*satilla, Adonis aestivalis, A. vernalis,*
*Alyssum montanum, A. calycinum,*
*Reseda lutea, Spiraea Filipendula,*
*Potentilla incana, P. rubens, Rosa*
*glauca, Astragalus cicer, A. danicus,*
*Hippocrepis comosa, Vicia tenuifolia,*
*Euphorbia Gerardiana, E. Cyparissias,*
*Helianthemum vulgare, Falcaria vul-*
*garis, Bupleurum falcatum, Salvia*
*pratensis, Calamintha Acinos, Stachys*
*rectus, Prunella grandiflora, Veronica*
*spicata, V. praecox, Asperula cynan-*
*chica, Aster Lynosyris, Artemisia cam-*
*pestris, Scorzonera purpurea.*

At Bamberg, an Association for
the protection of alpine plants has been
formed ; and a society for the preservation of the natural beauties of
the Isar valley, near München, has
been constituted.   Artists and brewers
more especially gave money to acquire

beautiful parts of this valley threatened
by industry.

We see, therefore, that the care
of natural monuments in Bavaria is
organised by government, which also
supports these efforts.

## *Hessen*

In the year 1902, a new law was
passed for the protection of monuments. It is remarkable that this
law comprehends not only monuments
of art and architecture, but also monuments of nature.

In 1904, the Department of Forestry
issued an order for the care of forests,
and of aesthetic and natural monuments. Notable rocks and trees must
be registered and photographed.

## Oldenburg

Two primeval forest reserves exist. In the forest district of Hasbruch, about 765 acres, and in the forest district of Neuenburg, about 1,185 acres, have been preserved. Both are well known, the last under the name of "Neuenburger Urwald." They are chiefly formed of beeches, hornbeams, and oaks ; but the following also occur :—*Alnus, Corylus Avellana, Prunus spinosa, Crataegus Oxyacantha, Ilex Aquifolium*, and *Rhamnus Frangula*. The oaks in particular are very large, as, for example, the "Thick Oak," the stem of which, at a height of 1 m, has a circumference of about 9 m, and the "Amalia Oak," the stem of which, at a height of 1·3 m, has a circumference of 9·4 m.

## *Prussia*

The organisation for the protection of natural monuments in Prussia was put in motion about ten years ago.

Like other naturalists, I also had observed that living nature was threatened on many sides by civilisation and cultivation, and that it was necessary to do something immediately to preserve as much as possible of what remained of primitive nature. I applied to the Department of Forestry as this department had control of considerable tracts of land, and would therefore be able to preserve natural monuments by administrative measures alone, without having to resort to special measures. In the beginning of the year 1899, I offered, to the chief of the Department of Forestry in Berlin, a memorial showing that the present existing method of managing the forests, especially the clear felling, resulted in the disappearance of virgin woodland and of indigenous plant and animal life. At first, I suggested that in all State forests the natural monuments, particularly remarkable portions of woods, rocks, fens, plant associations, groups of trees and herbaceous plants of scientific interest, and breeding places of rare birds, should be registered, mapped, and protected. At the same time, I submitted a record, with illustrations of the remarkable woods, trees, and shrubs of a single province, to support my appeal. I found the Department of Forestry very sympathetic, and it agreed to my proposals. The minister ordered the publication of the record, called the forest note-book (Forstbotanisches Merkbuch, see footnote, page 39), and distributed copies of it to each of the foresters of the

province, of whom there are about
400.

About the same time, W. Wete-
kamp, of Breslau, member of the
Prussian chamber of deputies, at the
end of 1898, showed that large sums
of money were being expended on
the maintenance of botanical gardens,
but that no funds were being employed
for the care of the natural flora and
fauna at home. He quoted several
cases of their being endangered, and
requested that the means for the
preservation of indigenous plant and
animal life should be supplied by the
government. The representative of
the Education Department agreed with
the deputy, and promised to give his
attention to the matter.

Later, in the year 1900, the Minister
of Education, together with the Minister
of Agriculture, Domains, and Forestry
requested reports on this subject from

certain geographers, geologists, botanists, and zoologists.   All reports laid
stress upon the importance of these
efforts, and advocated some kind of
State organisation for the protection
of natural monuments.  I was then en-
trusted by the Minister of Education
with the drawing up of a detailed
memorial (Denkschrift) setting out
distinct proposals for such an organisation.  This memoir was finished
early in 1904, and published by order
of the government[1].  It contains chiefly
the following proposals:—first, that
registers, maps, and illustrations should
be made of natural monuments of all
parts of the country; secondly, that
these natural monuments should be
protected; thirdly, that they should

[1] Conwentz, H.  Die Gefährdung der Naturdenkmäler und Vorschläge zu ihrer Erhaltung.
Denkschrift, dem Herrn Minister der geistlichen,
Unterrichts- und Medizinal-Angelegenheiten über-
reicht.  Berlin, 1904.—III. Auflage, Berlin, 1905.

be made generally known[1] (see page 36 *sq.*); and lastly that the government should assist and promote these efforts by establishing a special office for the care of natural monuments, and by instructing the officials of all departments actively to co-operate. The protection itself might be realised either by voluntary, by administrative, or by legislative help[2].

After this short historical introduction, a review of the chief measures

[1] Recently two publications of illustrations of natural monuments have been issued : Meerwarth, H. Lebensbilder aus der Tierwelt. Leipzig, 1908 ;— Schultz, G. E. F. Natur-Urkunden. Berlin, 1908. The latter book contains excellent photographs of plant associations, and of animals, taken from life.

[2] The British periodical "Nature" has already several times directed attention to the care of natural monuments in Prussia (see nos. 1830, 1926, 1962, 1978), and commented on those proposals as follows : "The various details in this proposed organisation for the protection of nature's monuments seem quite reasonable and eminently practical."

taken in this direction may very properly follow.

### *Parliament*

In 1902, a law was passed prohibiting the disfigurement of the country by advertisements. In 1907, another law was passed by which the local authorities were authorised to forbid that places of architectural or natural beauty should become defaced in any way. It is most undesirable that places of natural beauty should be openly disfigured by glaring advertisements, or by unsuitable and unattractive buildings. This observation applies, I am sorry to say, not only to Prussia and all Germany but to all civilised countries; and therefore other countries would be benefitted by the enactment of such laws. In England, a law forbidding the disfiguring of the landscape by advertisements is a matter of urgent necessity.

In 1906, parliament granted for the first time an annual sum of money for promoting interest in protecting natural monuments in Prussia. This money is not to be utilised for the purchase of lands possessing natural monuments, but only for the purpose of organising movements devoted to that end.

### Government

EDUCATION DEPARTMENT. In accordance with the above mentioned proposals, a "State Office for the Care of Natural Monuments in Prussia" was instituted, in the year 1906, under the superintendence of the minister; and a Prussian State Commissioner was nominated. The duty of the Commissioner is to schedule and collect information concerning natural monuments, and to determine upon measures for preserving them, if possible, by influencing the owners in favour of their preservation. The office has no financial resources for purchasing lands bearing natural monuments, which work is left to public bodies and societies. The Commissioner is prepared to answer every inquiry relating to the subject, and in case of necessity he goes to the spot. He has to make an annual report to the minister, and two reports have already been published[1].

In co-operation with the Prussian State Office, in each province a "Provincial Committee for the care of Natural Monuments"; and in districts, if it is possible, sub-committees are to

[1] Beiträge zur Naturdenkmalpflege, herausgegeben von H. Conwentz.—Heft I. Bericht über die staatliche Naturdenkmalpflege in Preussen im Jahre 1906. Mit 7 Abbildungen. Berlin, 1907.— Heft II. Bericht im Jahre 1907. Mit 5 Abbildungen. Berlin, 1908.

be formed for promoting greater interest in the country and for supporting the work of the State Office. The highest officer of the province (Oberpräsident) or of the district (Regierungspräsident) shall be in the chair, and a naturalist is also to be a member of each committee. At the present time, six provincial and six other committees are in existence. All these committees receive much support from provincial and municipal authorities. In several cases, special reports (Mitteilungen) are published by the committee.

The Minister of Education distributed many hundred copies of the "Denkschrift" and of the annual reports of the State Commissioner; and he invited all institutions of his department to assist the work of the Commissioner. For instance, all universities, academies, and schools of every kind were to have their interest awakened in favour of the care of natural monuments. The lecturers and professors of geography, geology, botany, and zoology were requested to keep these ideas in mind during their lectures. Moreover, at teachers' conferences the matter was discussed, and sometimes the State Commissioner has been present at such gatherings.

Again, the clergy were invited to assist in promoting this love of nature, and to inform the committee if a natural monument were threatened in any way.

Following these precedents, other departments have also interested themselves in furthering these efforts.

THE DEPARTMENT OF AGRICULTURE AND DOMAINS. This department has ordered that, in the forming or rearrangement of estates,

the preservation of beauties or rarities
of nature should receive more consideration than before. For instance,
brooks and rivers should be allowed to
flow in their original bed as far as
possible : new landmarks should be
so constructed as to accommodate
themselves to the old, as the latter
may bear remarkable trees or hedges,
characteristic groups of trees, rare
plant associations, and habitats of
interesting species; and hills, rocks,
and boulders should be saved by
being given over to the charge of
communities. In special cases, ground
of scientific value, such as a small
tract of moorland, should be expro-
priated for preserving it as a natural
monument. In general, the officials
should remember that it is their duty
not only to advance material ends,
but also to promote ideal aims. Then,
too, all rarities of the soil, or of plant

and animal life should be registered, and the lists should be subject to frequent revision. Moreover, the Minister suggested to the agricultural and veterinary academies that in lectures on geology, botany, and zoology reference should be made to the necessity of caring for and protecting natural monuments. Besides this, the State Commissioner was instructed to give special lectures on the subject at the various academies.

Some examples of realised protection of natural monuments by this department may be mentioned.

The sea-holly (*Eryngium maritimum*) on the sea-shore of East Prussia, West Prussia, and Pomerania is guarded by police orders. Some years ago, it was interdicted to uproot, or to offer the plant for sale, in order that this beautiful indigenous decoration of our sea-coast might be preserved.

On Memmert, an uninhabited
island between Juist and Borkum in
the German Ocean (figure 9), a sanctuary of birds has been instituted.
A watcher is placed there during the
breeding time, and this year about 300
pairs of the herring gull (*Larus argen-
tatus*), 600 pairs of terns—the common
tern (*Sterna hirundo*), the Arctic tern
(*S. macrura*), the lesser tern (*S. mi-
nuta*), and the sandwich tern (*S. can-
tiaca*), 30 pairs of the oyster catcher
(*Haematopus ostralegus*), and 50 pairs
of the Kentish plover (*Charadrius
alexandrinus*) were nesting there.

A tract of salt marsh, near Artern,
with the following characteristic salt-
marsh plants, was threatened by agriculture:—*Ruppia rostellata, Triglochin
maritimum, Cladium Mariscus, Atri-
plex pedunculata, A. littoralis, Sali-
cornia herbacea, Suaeda maritima,
Spergularia salina, Glaux maritima,*

Figure 9. Memmert Island, a Government Sanctuary
of Birds.

*Plantago maritima*, *Aster Tripolium*, *Artemisia maritima* var. *salina*. At the suggestion of botanists, it was decided to preserve this plant association in its pristine condition.

THE DEPARTMENT OF FORESTRY ordered that the proposals of the "Denkschrift" (see page 120, footnote) should be applied to the State Forests as far as possible. Clear felling is to be reduced, particularly near towns and watering-places. Boulders and remarkable rocks, small bogs, rare plants and plant associations, and characteristic specimens of trees are to be preserved. Parts of woods, which show indigenous or interesting vegetation, are to be protected. Various harmless birds and mammals are also to be protected.

In this sense, there are already a good many reservations in Prussian State Forests. For example, in the forest-district of Mirchau near Danzig, a beautiful example of woodland-scenery with lakes has been preserved, by complete clearing being prohibited. In the forest-district of Buchberg, a habitat of the Lady's slipper is enclosed with a hedge ; and in Wilhelmsberg, a Scots pine bearing a variety of the mistletoe (*Viscum album* var. *laxum*) is protected by a surrounding copse which has also been excluded from complete clearing. In the forest-district of Lindenbusch, an indigenous wood of 45·7 acres, the so-called "Ziesbusch," which means yew wood, is reserved. The wood contains many kinds of trees and shrubs, such as *Taxus baccata*, *Pinus sylvestris*, *Salix* spp., *Populus tremula*, *Corylus Avellana*, *Carpinus Betulus*, *Betula* spp., *Alnus glutinosa*, *Quercus Robur*, *Ulmus campestris*, *Crataegus* spp., *Pyrus communis*, *P. Aucuparia*, *Euonymus verrucosa*, *Acer plata-*

*noides, A. pseudo-Platanus, Rhamnus Frangula, Tilia* spp., *Hedera Helix, Cornus sanguinea, Fraxinus excelsior,* and *Viburnum Opulus.* The locality is very remarkable because of the underwood of more than 2,000 yews of different sizes, although the yew, in most of its stations, is a diminishing species. At Neulinum, close by the forest-district of Drewenzwald, a tract of moorland with the rare dwarf birch (*Betula nana*) was purchased by the Department of Forestry, in order that this interesting plant might be preserved as a natural monument. This fen-reservation contains, among others, the following vascular plants:—*Aspidium Thelypteris, A. cristatum, Nardus stricta, Calamagrostis neglecta, Aira caespitosa, Molinia coerulea, Eriophorum vaginatum, E. latifolium, E. angustifolium, Carex stricta, C. Goodenowii, C. panicea, C. acutiformis,*

*C. lasiocarpa, Salix caprea, S. aurita, S. myrtilloides, Betula verrucosa, B. nana, B. intermedia, B. pubescens, Drosera rotundifolia, Comarum palustre, Peucedanum palustre, Ledum palustre, Vaccinium uliginosum, V. Oxycoccus, Andromeda Polifolia, Calluna vulgaris, Lysimachia thrysiflora, Menyanthes trifoliata, Utricularia intermedia,* and *U. minor.* The bog-mosses are abundantly covered with many *Collembola* ; and the tussocks of grasses and sedges show the great abundance of certain members of the animal world, ants, caterpillars, spiders (e.g., *Agalena labyrinthica*), myriapods, and numerous others not yet fully investigated. On the dwarf birch, *Mytilespis, Amphigerontia intermedia,* and other Psocidae (or death-watches) are found.

In the forest-district of Chorin, near Berlin, some fenland with a lake, altogether more than 350 acres, have been reserved. A characteristic plant association has thus been safe-guarded. The following is a characteristic list of the species inhabiting this primitive piece of fenland :—*Aspidium cristatum, A. spinulosum, Equisetum pratense, Sparganium ramosum, S. minimum, Potamogeton gramineus, Scheuchzeria palustris, Sagittaria sagittifolia, Milium effusum, Holcus mollis, Brachypodium pinnatum, B. sylvaticum, Rhynchospora alba, Scirpus pauciflorus, S. sylvaticus, Eriophorum vaginatum, Carex* spp., *Caltha palustris, Juncus squarrosus, J. alpinus, Iris pseudacorus, Orchis latifolia, O. incarnata, Epipactis palustris, Malaxis paludosa, Salix pentandra, S. repens, Viscum album* on Birch, *Dianthus* spp., *Nuphar luteum, Nymphaea alba, Anemone Hepatica, A. nemorosa, Ranunculus aquatilis,*

*R. Lingua, Drosera* spp., *Chrysosplenium alternifolium, Parnassia palustris, Ribes* spp., *Prunus spinosa, Spiraea Filipendula, Comarum palustre, Trifolium alpestre, T. fragiferum, Astragalus glycyphyllus, Vicia* spp., *Lathyrus montanus, Callitriche stagnalis, Impatiens noli-tangere, Hypericum* spp., *Viola palustris, V. hirta, V. sylvatica, Epilobium roseum, E. palustre, Circaea lutetiana, C. alpina, Cicuta virosa, Sium latifolium, Selinum carvifolia, Peucedanum palustre, Ledum palustre, Vaccinium Oxycoccus, Andromeda Polifolia, Hottonia palustris, Menyanthes trifoliata, Myosotis versicolor, Veronica scutellata, Pedicularis palustris, Melampyrum pratense, Utricularia vulgaris, U. minor, Galium* spp., *Campanula rapunculoides, Eupatorium cannabinum, Solidago Virgaurea, Gnaphalium sylvaticum, Bidens tripartita,*

*Carlina vulgaris, Carduus nutans, Cnicus* spp., *Arnoseris minima, Lactuca muralis,* and *Crepis paludosa* (after E. Ulbrich).

In the forest-district of Thale in the Hartz Mountains, one of the prettiest mountain valleys in Germany, the "Bodetal" is safe-guarded as a monument of nature, and all projects of constructing a railway to the "Rosstrappe" and "Hexentanzplatz" have been suppressed (see page 6). Moreover, the woods growing spontaneously on both sides are preserved. In the forest-districts of Colbitz and Planken, a wood of lime-trees, formerly occupying about 988 acres, is protected.

In the forest-district of Münster, a valuable wood of oaks, beeches, hornbeams, ashes, and other trees is not to be cut down, in order that certain rare lichens, which are growing on the bark

of those trees, may be preserved. Thus a permanent object of education and study to the neighbouring university is secured. The most important species of lichens which occur there are :—
*Arthopyrenia antecellens, A. cinereopruinosa, Acrocordia tersa, Sagedia olivacea, Pyrenula laevigata, Calicium adspersum, C. lenticulare, Sphinctrina tubaeformis, S. turbinata, Arthonia fuliginosa, A. gregaria, A. impolita, A. marmorata, A. medusula, A. obscura, A. pineti, A. stellaris, Opegrapha Turneri, Graphis elegans, G. serpentina, G. dendritica* forma *acuta, G. conglobata, G. stellaris, Enterographa crassa, Lecanactis abietina, L. illecebrosa, L. lyncea, Pertusaria leptospora, P. multipuncta, P. pustulata, Thelotrema lepadinum, Gyalecta Flotowii, Phialopsis rubra, Biatora diluta, B. Lightfootii, B. quernea, Bacidia albescens, B. carneola, B. Friesiana, Bilim-*

*bia sphaeroïdes, Scoliciosporum leci-
deoïdes, Buellia arthonioïdes* (Zopf).

In the forest-district of Wolfgang,
29 beautiful old oaks and beeches are
reserved for the students of the
drawing academy at Hanau.

It was also ordered that the natural
monuments of each forest-district
should be registered by the forest-
office, and that official records of natural
monuments should be instituted by
each forest-office. Copies of these
registers must be sent to the State
Office for the care of natural monuments, every time an entry is made. It
was further recommended that in each
province a forest note-book should be
drawn up and published. The one published for West Prussia (see page 39),
nine years ago, has been followed by
others (Hanover, Pomerania, Slesvic-
Holstein) edited and distributed by the
Department of Forestry. Again, the

office for forest management (Forstein-richtungs-Bureau) has been charged with inserting the position of remarkable natural monuments upon the forest-maps, and a great number of these new maps has been already printed.

I was requested by the Minister to deliver lectures at the Prussian forest academies, on the preservation of natural monuments in the forests. I gave these special lectures at the Academies of Forestry, and also at the Academies of Agriculture and Veterinary Science, in the years 1907 and 1908.

THE DEPARTMENT OF PUBLIC WORKS instructed its engineers and architects to direct their attention to the preservation of natural monuments. In constructing channels, roads, railways, or buildings, it should be their care that essential monuments of nature should not be threatened or violated. In all doubtful cases, the advice of the State Office was to be solicited.

The administration of railways in constructing new lines and in enlarging stations (Ehrenbreitenstein) sometimes has regard to the protection of the beauties of scenery. Moreover, the administration occasionally incurs considerable expense in preserving old and remarkable trees, as at Barmen, Dortmund and Sedlinen. It was also recommended that railway embankments should be planted with shrubs, in which birds may build their nests.

The Home Department directed the chairmen of the land districts (Landräte) and the magistrates of the towns to promote the interests of the whole population by protecting natural monuments. This requirement for the care of natural monuments must only be ignored, when inconsistent with the interests of the public health.

THE WAR OFFICE commanded that on military lands and waters a register should be made of natural monuments that they may be respected; and a copy of the register must be sent to the State Office. Further, these natural monuments are to be protected as much as possible. Two examples may be mentioned here. The exercise ground near Halle encloses a small moor with an interesting vegetation, which was threatened by cultivation. It was commanded that the land should remain unchanged, for the purpose of study. Again, in the neighbourhood of Fort Weichselmünde, near the mouth of the Vistula, a habitat of *Eryngium campestre*, which has its eastern limit at this point, has been preserved.

THE TOPOGRAPHICAL DEPARTMENT of the General Staff has ordered that certain natural and prehistoric monuments are to be noted, and their positions marked on the maps. A list of these monuments is to be prepared by the State Office every year.

## Public Corporations

Various public bodies have followed the example set by the government, and others have taken up the subject. The province of Rhenish Prussia granted £10,000, and the corporations of Cologne and Bonn £5,000 and £2,500 respectively for protecting the Siebengebirge against spoliation. The province of Slesvic-Holstein paid £40, and a district (Sonderburg) paid £32, for preserving a large boulder. The administrative body of a West Prussian land district (Karthaus) bought land containing a remarkable moraine, and thus preserved it. The corporation of Danzig made a reservation in one of the town forests for protecting a small lake with a rare variety of fish (*Phoxinus*

*laevis* var. *punctatus*), and a moor with the following characteristic plants:—
*Aspidium Thelypteris, A. cristatum, Scheuchzeria palustris, Eriophorum* spp., *Salix aurita, Comarum palustre, Rhamnus Frangula, Hydrocotyle vulgaris, Peucedanum palustre, Ledum palustre, Vaccinium uliginosum, V. Oxycoccus, Andromeda Polifolia, Lysimachia vulgaris, Menyanthes trifoliata.* The corporation of Frankfort-on-the-Main made a forest reservation of 72·5 acres for the purpose of instruction and study. There are many other towns and communities in addition to those mentioned which have made land reservations of a similar kind.

### Societies

The care of natural monuments is quoted as an aim in the rules of many associations, for instance, the German

Association of Systematic Botanists
and Plant Geographers, the German
Teachers' Union for Natural History,
the West Prussian Botanic and Zoological Society, the Natural History
Society at Geestemünde on the Weser,
and the Berlin Society for Promoting
Knowledge of the Homeland.

Several societies have made grants
of money for purchasing natural monuments. The Travellers' Association
at Elbing bought some land near the
coast with beeches growing on it, close
to the eastern geographical limit of that
tree, for preserving it in its natural
state. The Anthropological and Ar-
chaeological Society at Guben saved
a big boulder in the same way. The
Aller Association at Neuhaldensleben
farmed some fenland for many years,
as it was not saleable. Thus a characteristic plant association was preserved
which contains the following mem-

bers:—*Ophioglossum vulgatum, Carex
Œderi, Iris sibirica, Salix pentandra,
S.nigricans,S.aurita × repens, Thesium
intermedium, Drosera rotundifolia,
Spiraea Filipendula, Comarum palus-
tre, Potentilla alba, Trifolium mon-
tanum, Orobus vernus, Gentiana
Pneumonanthe, Cuscuta epithymum,
Pinguicula vulgaris, Arnica montana,
Scorzonera humilis.*

Some economic societies have also
supported these efforts, such as the
Silesian Fishery Society, which abol-
ished the premium for killing the
halcyon (*Alcedo hispida*). The West
Prussian Union for the protection of
game voted that its members should
not shoot or catch the owl, eagle owl,
raven, sea eagle, golden eagle, and
black stork.

### *Individuals*

Private individuals support these efforts very often. In the first place, there are many landowners who pre- serve natural beauty, remarkable rocks, woodland scenery, rare trees, and nests of birds on their own estates. Thus, Prince Stolberg-Wernigerode prohi bited the building of a theatre at the Brocken, so as to preserve the natural beauty of the highest summit of the Hartz mountains. Prince Putbus re- served the indigenous wood of the isle of Vilm near Rügen ; and Count Dohna Finckenstein reserved about 88·9 acres of his forests, specially for the protection of the natural landscape around the lakes.

Other persons have granted money for preserving natural monu- ments by purchase. For example,

Professor Thomsen bought 51·87 acres
of heathland near Mount Wilsede in
the Lüneburger Heide, for preserving it. Professor Geisenheyner of
Kreuznach collected money for purchasing a place with a pontic plant
association near Waldböckelheim in
the Nahe district[1]. This vegetation
consists of :—*Asplenium Ceterach, A.
Trichomanes, A. septentrionale, A.
Ruta-muraria, Stipa pennata, Avena
pratensis, Aira caryophyllea, Poa
bulbosa, Carex humilis, Gagea pratensis, Allium sphaerocephalum, Dianthus Carthusianorum, Alsine tenuifolia, Cerastium semidecandrum* var.
*glutinosum, Alyssum montanum,
Reseda luteola, Sedum album, S.
acre, S. reflexum, Fragaria viridis,*

---

[1] Geisenheyner, L. Ueber Naturdenkmäler,
besonders im Nahegebiet. *Allg. Bot. Zeitschr.
für Systematik, Floristik, Pflanzengeographie u.s.w.*
Jahrg. 1904, Nr. 10 und 11. Karlsruhe, 1904.

*Potentilla rupestris, P. arenaria, Rosa pimpinellifolia, Genista pilosa, Ononis repens, Medicago minima, Trifolium alpestre, T. rubens, Oxytropis pilosa, Lathyrus niger, Geranium columbinum, G. rotundifolium, Linum tenuifolium, Acer monspessulanum, Eryngium campestre, Carum Bulbocastanum, Bupleurum falcatum, Ligustrum vulgare, Cynoglossum officinale, Origanum vulgare, Galeopsis angustifolia, Stachys rectus, Teucrium Botrys, VerbascumLychnitis,VeronicaDillenii, Orobanche lutea, O. alba, Asperula glauca, A. cynanchica, Aster Lynosyris, Erigeron acre, Anthemis tinctoria, Achillea nobilis, Artemisia campestris* var. *sericea.*

Sometimes the protection of a natural monument by purchase cannot be effected by a community, or society, or an individual alone, but only by the co-operation of various organisations.

The preservation of a very interesting moorland of 4·12 acres at Schafwedel, near Uelzen, Hanover, was brought about in this manner. On this moorland, the following plants occur:— *Aspidium Thelypteris, Pinus sylvestris, Juniperus communis, Agrostis canina, Briza media, Festuca elatior, F. ovina* var. *capillata, Carex rostrata, Luzula campestris, Salix pentandra, S. repens, Populus tremula, Betula verrucosa, B. pubescens, B. alpestris, B. nana, Caltha palustris, Drosera rotundifolia, D. anglica, Parnassia palustris, Comarum palustre, Potentilla sylvestris, Linum catharticum, Radiola linoides, Empetrum nigrum, Hypericum acutum, Lythrum Salicaria, Epilobium palustre, Hydrocotyle vulgaris, Vaccinium Oxycoccus, Calluna vulgaris, Erica Tetralix, Menyanthes trifoliata, Lycopus europaeus, Euphrasia stricta, Galium uliginosum,*

*Crepis paludosa, Hieracium laevigatum.*

The full amount for purchasing this moorland was subscribed by public corporations, several natural history and teachers' societies, and private individuals. The preservation of this moor is extremely valuable, as in the whole flat land of North Germany only two habitats of *Betula nana* are known, namely, Neulinum in West Prussia, and Schafwedel in Hanover. These are now preserved, the first by the Department of Forestry and the other by private help.

Some German artists at Rome bought the oak wood near Olevano in the Sabine mountains, celebrated in verse by v. Scheffel, and offered it to the German Emperor. He accepted the present, not for himself but for the German Empire.

In Prussia, then, the care of

natural monuments is organised by government, and promoted by parliament, by communities, by societies and by individuals.

## Saxony

At the beginning of 1904, at a joint meeting of the Geographical and the Natural History Societies of Dresden, I lectured on the protection of nature's monuments, with special reference to Saxony. Government delegates and professors of the Technical Academy at Dresden, of the Academy of Forestry at Tharand, and of the Mining College at Freiberg were present. Later, a General Committee for the preservation of natural monuments was formed, and it is now supported by the government.

Before this, the government of Saxony had bought a forest, threatened

by clearing operations, in Saxon Switzerland, near the Bastei, thus preserving
a favourite pleasure resort. Further, in
celebration of the King's jubilee, the
corporation of Dresden purchased a
forest of 286 acres to preserve it, as
a natural monument, in its entirety.
A boulder with glacial striae has been
preserved for thirty years by the
Humboldt Society at Löbau.

### Reichsland

Several years ago, a committee
for the protection of natural monuments was formed by the Philomatic
Society at Strassburg.

### German Colonies

Some of the German colonies are
beginning to care for natural monuments, and a few instances may very
well be quoted here.

EAST AFRICA.—A part of the Rain Forest in East Usumbara is given over to the care of the Agricultural Biological Institute of Amani. The main trees of that forest are: *Allanblackia Stuhlmannii, Brochoneura usumbarensis, Chlorophora excelsa, Chrysophyllum msolo, Cyathea Manniana, Mesogyne insignis, Myrianthus arborea, Sorindeia usumbarensis* (Volkens). Other Forest Reserves have been made in West Usumbara, where *Agauria salicifolia, Juniperus procera* and *Podocarpus usumbarensis* are growing; and in the Sachsenwald, near Dar-es-Salam, where the woods contain *Bombax rhodognaphalon, Cardiogyne africana, Erythrophloeum guinense, Pterocarpus erinaceus*, and *Trachylobium verrucosum*.

KAMERUN.—Some parts of the forests of the Kamerun Mountains are protected. Remarkable species, such as:—*Ceiba pentandra, Kleinedoxa Zenkeri, Carpodinus landolphioides, Mimusops djawa*, and *Spathodea campanulata* are there indigenous.

South West Africa.—It is prohibited, by a regulation of 1902, to catch or to kill certain mammals, such as the elephant, the hippopotamus, the giraffe, and the zebra, and such birds as the ostrich and the topau. Exceptional treatment is, of course, given to scientific expeditions.

In 1907, the habitat of *Welwitschia mirabilis*, near the station Welwitsch, on the Windhuk line, was enclosed by means of a fence.

Lastly, it may be mentioned that a Committee has been formed in Berlin, which aims at preventing the complete extermination of the African elephant.

## Holland

On the invitation of the Dutch Botanical Association, in 1904, I read a paper on the protection of remarkable plants and animals[1]. In the following year, at Amsterdam, a conference took place for the purpose of discussing the steps which should be taken for the protection of natural monuments, and an Association for the Protection of Natural Monuments in Holland was instituted.

The reservation of the Naardermeer, in the south of the Zuider Zee, is of the greatest importance, as all lakes and moors there are threatened by draining. The avifauna of the Naardermeer is of very great interest.

[1] Nederlandsch Kruidkundig Archief. Nijmegen, 1904, p. 91 sq.

## *Dutch Colonies*

In the island of Java, in the Dutch East Indies, there is a mountain forest, near Tjibodas, and on the north-west side of the Gedeh Mountains, at a height of 1,392 to 1,787 m, belonging to the Botanical Garden of Buitenzorg. In 1889, this primeval forest of nearly 700 acres was converted by the government into a reservation. It was taken charge of by the Botanical Garden.

## Russia

In the Bjelowesh Forest, the European bison (*Bison europaeus*) is completely protected by law.

By private help, too, a natural monument of great value has been saved. In the south of Russia, in the Taurine District, a German-Russian landlord made a steppe-reservation of about 247 acres for protecting the interesting vegetation (*Stipa pennata, St. capillata, Paeonia tenuifolia, Prunus chamaecerasus, Salvia nutans, S. pratensis*), and the animal world too (Saiga).

## Sweden

A. E. Nordenskiöld, the well-known investigator of the Polar regions, advocated, in 1880, that natural reserves ought to be instituted not only in Europe but in America as well. No harm could be done by reserving an original forest of considerable extension in the north of Sweden for instance, and future generations would profit by its study. J. Gunnar Andersson, now Director of the Swedish Geological Survey, proposed to reserve Gotska Sandön; but the time was not ripe for such ideas, and these wise proposals were passed over without notice.

Again, in 1903, I received an invitation to lecture on the protection of natural scenery, of its original vegetation and animal world, with special reference to Sweden. These lectures took place at Stockholm, Upsala, Göteborg, and Lund, in January, 1904; and the paper was published by the Swedish Society for Anthropology and Geography[1].

Then, in the same year, a bill for the preservation of natural monuments was brought in by Lektor K. Starbäck, member of the Riksdag, and the government requested a memorial from the Royal Academy of Sciences. In this memorial, protection by law

[1] Conwentz, H. Om skydd åt det naturliga landskapet jämte dess växt-och djurvärld, särskildt i Sverige. Med 7 illustrationer. Ymer. Årg. 1904, pp. 17–42.

was proposed of the puffin (*Fratercula arctica*), and the government decided in favour of the proposal. Further, induced by the Academy, the small lake of Fagertärn, in the province of Nericke, where a beautiful water-lily (*Nymphaea rosea*) grows, and part of another lake in Scania, Immelnsjö, with the only living remains of water nut (*Trapa natans*) have been protected. It remains to be mentioned that two persons, who have taken an active part in the protection of natural monuments, have been distinguished by medals of the Academy of Sciences.

Last year, the Department of Agriculture appointed a Committee for the protection of the more remarkable and interesting natural objects of the country. This Committee consists of Häradshöfding E. L. Améen, Professor Einar Lönnberg, and Lektor K. Starbäck. At the end of 1907, the Committee delivered to the Department of Agriculture several bills concerning these questions. One of these refers to the reservation of National Parks and their management. Several such National Parks are proposed in different parts of Sweden. A second bill refers to the reservation of smaller areas for the protection of certain plants, animals, and other natural monuments. A third bill proposes that in certain instances landed property should be appropriated for public purposes and for the protection of indigenous natural objects. If the bills are not rejected by the Riksdag, these matters will be successfully arranged. It is expected that when the laws mentioned have been passed a great deal will be done spontaneously by the owners of real estates and by gratuitous gifts, as some smaller areas

have indeed already been promised
(Lönnberg).

Natural history and geographical
societies at Stockholm, Upsala, and
Lund have discussed and promoted
the subject of protection of natural
monuments. For example, " Lunds
Botaniska Förening" requested all
botanists to be careful with rare
plants, and it excluded the following
rare species from its exchange list :—
*Scolopendrium vulgare, Equisetum
maximum, Carex punctata, Allium
carinatum, Genista germanica, Astragalus danicus, Trapa natans, Primula
sibirica*, and others. Again, the Skogs-
vårdsförening, in Stockholm, proposed
that on the Bonden Island, in the Gulf
of Bothnia, the destruction of the
razor-bill (*Alca torda*) should be prohibited; and the government agreed
to the proposal.

## Switzerland

In all cantons, laws exist for the protection of the Edelweiss (*Gnaphalium Leontopodium*). In the Canton of Schaffhausen, the rare plants are protected by a forest-law of 1904 ; and in the Canton of Valais, there is an interdiction of 1906 to uproot, to expose to sale, and to send off alpine plants, medicinal ones alone being excepted.

In other cases, natural monuments are guarded by administrative measures. The community of Oensingen preserves the unique Swiss habitat of *Iberis saxatilis*; and, near the Beatenhöhle, which is much frequented by travellers, *Cyclamen europaeum* is preserved.

In 1883, the "Association pour la protection des plantes," which publishes annual reports and which also aims to establish alpine gardens, was formed

11—2

at Geneva.   The local natural history
societies have been anxious to preserve
the boulders in Switzerland, and one
of these societies (St Gallen) has
about 150 in its possession.   In 1906,
a General Committee for the protection
of natural and prehistoric monuments
was formed by the Swiss Natural
History Society.   It shows great activity, and publishes particulars every
year.   Committees, with the same
objects, exist in the cantons.   The
president of the General Committee,
Dr P. Sarasin, and his cousin Dr J.
Sarasin, had, six years ago, preserved
a piece of land containing a large and
beautiful yew-tree, near Heimiswyl, by
actual purchase.

## The United States of America

PARLIAMENT. The idea of preserving primeval nature in the United States has been realised in a veritably grand manner. It is, however, true that formerly vast districts of territory had been deprived of their primitive vegetable and animal world, in great part or almost entirely. Six various landscapes, distinguished by natural or prehistoric monuments, and some of them very extensive, have been reserved for the public by Congress (figure 10). It is strictly prohibited to shoot, to catch, to injure, or to kill a bird or any other wild animal there, unless it is those which attempt one's life.

The Yellowstone National Park,

Figure 10.   National Parks of the United States
of America.

established in the year 1872, is the oldest and largest one. It is chiefly situated in the State of Wyoming, and covers a space of 3,348 square miles. The park contains numerous falls and rapids, picturesque rocks, fine forests, and, above all, about 200 head of the American bison (*Bison americanus*) which in general is becoming extinct.

The Sequoia National Park, in California, was instituted in 1890, particularly for the protection of the American Big Tree (*Sequoia gigantea*). This reservation has an extent of 250 square miles. Some of the trees attain a height of more than 300 feet; and sometimes the bases of the trunks are so thick that gateways for carriages have been made through them. The largest trees of the kind are labelled with the names of celebrated persons.

The General Grant National Park, also in California, was instituted in the same year. The park occupies four square miles.

The Yosemite National Park, in California, was founded in 1890. The district occupies 1,512 square miles, and contains some of the most beautiful parts of the Sierra Nevada.

The Mount Rainier National Park, in the State of Washington, is about nine square miles in extent, and was instituted in the year 1899. It contains glaciers and a great many Arctic plants and animals. Although of great altitude, rare birds and other animals were threatened by sport, and it was on this account that the reservation was made.

The Arizona National Park, near Holbrook, in Apache County, was instituted in 1906. It forms a so-called petrified forest, consisting of silicified stems of trees which appear to resemble those of existing Araucaria.

Some of the trunks have a length
of 100 to 200 feet, and a thick-
ness of seven to ten feet. Even
in prehistoric times, stone-hammers,
arrow-heads, and knives were made
of the petrified wood, whose remains
are still to be found. Again, old
Indian houses are built of this beautiful material. In recent times, the
material has been exploited by industry; and several companies on the spot
manufactured objects of art and ornaments from it. A great many of the
latter were sold at the Paris Exhibition,
in 1900. Thus the existence of the
silicified forest was endangered, and a
Public Reserve was established, on the
suggestion of the State of Arizona.

ASSOCIATIONS. There are many
scientific and other societies working
at the investigation and protection
of natural monuments in the U.S.A.
The American Scenic and Historic

Preservation Society, founded in 1895, is a national association, for the purpose of preserving natural scenery of scientific or aesthetic interest. The Wild Flower Preservation Society of America was founded in 1902. The organisation of this national society is the direct result of an essay by Dr Knowlton, which was awarded the first prize in the competition held by the New York Botanical Garden with the income of the Caroline and Olivia Phelps Stokes Fund[1]. The society edits a monthly journal called "The Plant World."

The National Association of Audubon Societies aims at preventing, so far as possible, the killing of any wild bird not used for food, the destruction of nests or eggs, and

[1] Knowlton, F. H. Suggestions for the preservation of our native plants. *Journal of the New York Botanical Garden.* Vol. III. 1902.

the wearing of feathers as ornaments for dress. The Association, founded in 1901, publishes annual reports and numerous pamphlets with illustrations.

The American Bison Society advocates the protection of the bison.

### Connecticut

In Connecticut, there is a Law of 1899 protecting wild flowers and trees:—"Every person who shall wilfully injure any tree or shrub standing upon the land of another, or on the public highway in front of said land, or injure or throw down any fence, trellis, framework, or structure, on the land of another, or shall wilfully cut, destroy, or take away from the land of another, any creeping fern, crops, shrub, fruit or vegetable production, shall be fined not more than one hundred dollars or imprisoned not more than twelve months, or both." The words regarding the destruction of "any creeping fern" were intended to protect the Hartford fern, the walking fern, and maiden hair.

Another paragraph reads thus : " Every person who shall wilfully pull up, tear up, dig up, or destroy any trailing Arbutus from the land of another, or who shall sell, expose for sale, purchase, or have in his possession, any Arbutus with the roots or underground stems attached, shall be fined not more than twenty dollars ; provided, however, that any person may take such Arbutus on land owned or leased by him, or with the permission of the owner or lessee."

### Massachusetts

In Massachusetts, by a law of 1891, Trustees of Public Reservations were

instituted for the purpose of acquiring
places of natural beauty or historic
interest. Hitherto a great many such
National Parks have been reserved, and
there is a local committee for each one.
In 1899, about 3,000 acres of land were
set aside on Wachusett Mountain as
a State Reservation, and the commissioners in charge were given police
powers. This should ensure a permanent game sanctuary for Worcester
County. The enactment of 1907, by
which the Commissioners on Fisheries
and Game were empowered to take
1,000 acres of land on Martha's Vine-
yard as a reservation for the protection
of the heath hen and other birds, is an
example of direct legislation for this
purpose, more of which will become
necessary[1].

[1] Forbush, E. H. Statutory Bird Protection in
Massachusetts. *Bulletin of Massachusetts State
Board of Agriculture.* 1907.

In other States of the U.S.A., similar State Parks and game preserves have been established.

In Boston, there is a Society for the protection of native plants, which publishes a bulletin.

A private donation too is of interest. Mrs Fanny Foster Tudor founded in remembrance of her late daughter, Virginia, the "Virginia Wood," which is a coniferous wood of twenty acres in Stoneham.

### International

The care of natural monuments is not only a national, but also an international affair ; and there are already beginnings of international organisation of the same sort. For example, at the first Peace Conference held at the Hague in 1899, the delegates agreed (art. LV.) as follows :

" The occupying State shall only be regarded as administrator and usufructuary of the public...forests and agricultural works belonging to the hostile state and situated in the occupied country. It must protect the capital of these properties and administer it according to the rules of usufruct." Thus, in time of war, the complete devastation of forests will henceforth be prevented as much as possible.

The International Convention, held in London in 1900, purposed to preserve the various forms of animal life existing in a wild state within a certain large zone of Africa. The hunting and destruction of the following animals are more particularly prohibited: firstly, on account of their usefulness, vultures, secretary bird, owls, and rhinoceros bird; secondly, on account of their rarity and threat-

ened extermination, giraffe, gorilla,
chimpanzee, mountain zebra, wild
asses, white tailed gnu, elands, and
little Liberian hippopotamus ; and
thirdly, the hunting and destruction of
any other animals whose protection,
whether owing to their usefulness or
to their rarity, may be considered
necessary by the local government is
prohibited in the same way. Moreover, reserves are to be established
within which it is unlawful to hunt,
capture, and kill any bird or other
wild animal except those which are
exempted from protection by the local
authorities.

The International Zoological Congress, held at Berlin in 1901, passed
a resolution in favour of the protection of all innocuous higher animals
threatened by the progress of cultivation. In the year 1905, the International Botanical Congress, held at

Vienna, presented to the Bosnian government a memorial asking for reserves in Bosnia in order that certain indigenous plant associations of great interest might be protected.

## SUGGESTIONS

All in all, the results of the movement in the United Kingdom and its
possessions are that not a few arrangements have already been made to
ensure the care of natural monuments,
for places of natural beauty, characteristic soil formations, plant associations,
habitats of plants, and breeding places
of birds have been preserved. As
regards the further promotion of these
efforts, procedure by government
would not, in general, appear to be
the right method in Britain. Speaking
generally, it is by self-help that Britain
has become great; and in this case
also self-help may be recommended for

promoting this matter of scientific interest and public benefit. No doubt, corporations and societies of various kinds will be willing to support voluntarily these endeavours. For instance, the National Trust will always endeavour to protect districts of scientific interest, and natural history societies might follow the example of the London Geological Society in granting money for preserving natural monuments. The Central Committee for the Study and Survey of British Vegetation carries out the investigation and mapping of British vegetation in an exquisitely methodical manner. I should like to suggest that this committee should add to its objects the protection of characteristic plant associations, and of single rare species. Besides, all other natural history associations, in making their regulations, should take into consideration not only the investigation but also the protection of indigenous nature.

In the future, it will be necessary to bring together and to organise all single efforts in the British Isles, and to establish a focus of activity. If I am right, the British Association for the Advancement of Science would be the proper place for this. The "B.A." comprehends all branches of natural science, and is an authority to the whole nation. At every meeting, a great many members and delegates of natural history societies and field clubs of the country are present. It would not be necessary to establish a new section for this matter; rather the subject would be gone into, now and then, in existing sections. Further, as there are already committees of geological and of botanical photographs, so might a committee for the protection of natural monuments be instituted, and assisted by delegates of the universities, of the great learned and other societies of the National Trust, of the Geological Survey, of the Committee for the Study and Survey of British Vegetation, and of county councils and other local authorities.

To begin with, the end and aim of the committee must be precisely defined. Above all, registers or inventories and maps of the natural monuments of all parts of the country should be made, and it should there be noted if a natural monument has been hitherto unprotected, and if it is threatened by destruction. Illustrations of natural monuments should be made if they do not exist already. Particularly, when in any given case there is no chance of preserving the object itself before its ruin, at all events it should be preserved by photography or other kind of illustration. Publications of these lists, illustrations, and maps are much to be desired, so long as the existence of the natural monument is not endangered by such publication.

There is no doubt that, if once the care of natural monuments were to come under the auspices of the British Association, it would receive considerable impulse from it, and a most laudable object would be constantly kept before the eyes of the people.

As regards the international protection of natural monuments, there is still much that remains to be done, especially in those countries which have not yet been appropriated. Certain fishes and other aquatic animals whose existence is threatened in some parts of the ocean, the Spitzbergen reindeer, and the musk-ox in Greenland, ought to be protected. When diplomatic treaties on the future of this northern island are made, the protection of the reindeer, which is rapidly diminishing, ought to be discussed and considered. Also in the Antarctic region, international arrangements are highly desirable. According to books of travel, the penguins and other animals of the Antarctic do not yet stand in awe of man ; and it will be a matter of great satisfaction, if, in this far-off district, nature's peace is preserved as long as possible.

Further, governments should strengthen the protection of the whole fauna all the world over, wherever the animals do not come into dangerous competition with the welfare of human beings. Sir H. H. Johnston, formerly Governor of Uganda in Central Africa, says (see page 33, footnote), " The world will become very uninteresting if man and his few domestic animals, together with the rat, mouse, and sparrow are its only inhabitants. Man's interests must come first, but those very interests demand food for the intellect. Aesthetically, the egret, toucan, bird of paradise, grebe, sable, chinchilla, and fur-seal are as important as the well-dressed woman. The viper, lion, tiger, crocodile, wolf, vulture, and rhinoceros have all their places to fill in our world picture. They are amazingly interesting, and therefore their destruction should only be carried out to the degree of keeping them in their proper sphere."

For some years past, periodic meetings of delegates of European, American, and Asiatic academies of sciences have been held, for deliberating on scientific affairs of general interest. Such international conferences afford a suitable opportunity for considering and supporting the protection of natural monuments in un-owned territories, and in various parts of the ocean. Such an international association of scientific academies might, without impropriety, make suggestions or recommendations to the States concerned, on such an uncontroversial matter as the preservation of natural monuments.

In conclusion, it may be pointed out that the care of natural monuments is not only of scientific and public interest, but it also possesses a patriotic value; for, by these undertakings, parts of the country at home become better known and more fully appreciated. In this way it is that true patriotism—the love of one's homeland—is increasingly promoted. True patriotism is one of the finest national characteristics of any civilised people; Shakespeare says:—"Who is here so vile that will not love his country" (*Julius Caesar*, Act III. Sc. II.).

Printed in the United States
By Bookmasters